BULK METALLIC
GLASSES AND
THEIR COMPOSITES

BULK METALLIC GLASSES AND THEIR COMPOSITES

Additive Manufacturing and Modeling and Simulation

MUHAMMAD MUSADDIQUE ALI RAFIQUE

MOMENTUM PRESS
ENGINEERING

Bulk Metallic Glasses and their Composites: Additive Manufacturing and Modeling and Simulation

First published in 2018 by
Momentum Press®, LLC
222 East 46th Street, New York, NY 10017
www.momentumpress.net

ISBN-13: 978-1-94708-384-4 (paperback)
ISBN-13: 978-1-94708-385-1 (e-book)

Momentum Press Emerging Materials Collection

Cover and interior design by S4Carlisle Publishing Service Ltd.
Chennai, India

First edition: 2018

10 9 8 7 6 5 4 3 2 1

Printed in the United States of America

Dedication

The author would like to dedicate this work to Abu Musa Jabir ibn Hayyan (c 721–c 815) (Father of Chemistry).

ABSTRACT

Bulk metallic glasses have emerged as competitive engineering material and have captured the attention of researchers across the globe because of their excellent mechanical properties (high hardness, high strength, and high elastic strain limit). However, they suffer from lack of ductility and fail catastrophically under tension. To this end, this problem can be overcome by forming a metal matrix composite such that some crystalline phases are introduced in the alloy during solidification, which provides a means of hindering rapid motion of shear bands. Thus, ductility and toughness increase while retaining high strength. The methods by which these crystalline phases are introduced and how they control the microstructure have come under intensive investigation over the years. Various mechanisms (such as *ex situ* introduction, *in situ* precipitation, or devitrification) are proposed on how to introduce crystalline phases and increase ductility and toughness. Recently, additive manufacturing has been proposed as the final solution of the problem as the final complex shape can be produced in a single step with composite structure in whole part exploiting the inherent nature of the process. However, this technique is still in its infancy and numerous challenges exist on how to produce the final part without defects and how to control final microstructure. This study is aimed to address this problem from solidification processing and modeling and simulation perspective. A comprehensive coupled macroscopic and microscopic model is proposed to predict microstructure of solidifying alloy in liquid melt pool of additive manufacturing. Microstructure control is exercised by introducing inoculants during solidification. Their number density, size, and distribution are hypothesized to control microstructure, and this is studied experimentally and validated by modeling and simulation. The methodology is claimed to be meritorious. The work is primarily intended for researchers, material scientists, doctoral candidates, practicing engineers, technologists and professors in academia, national laboratories, and industry.

KEYWORDS

additive, bulk metallic glasses, composites, modeling and simulation, solidification

CONTENTS

ACKNOWLEDGMENTS

Author would like to acknowledge the motivation and support of Prof. Milan Brandt throughout his PhD studies candidature. He also likes to thank RMIT University for providing scholarship for tuition and living and administration team (especially Mrs. Lina Bubic) of the School of Engineering for being extra supportive and encouraging. He also gratefully acknowledges the experimental support provided by CSIRO for this work.

RESEARCH SIGNIFICANCE AND BACKGROUND

Discovered in 1960 by Duwartz et al. (Klement et al. 1960) at Caltech, metallic glasses emerged as a completely new class of materials exhibiting very high tensile strength, hardness, elastic strain limit, and yield strength at relatively lower density as compared to steel and other high-strength alloys (Telford 2004; Schuh et al. 2007; Inoue and Takeuchi 2011). Yet, their use has not been able to get broad acceptance as competing engineering material because of the lack of ductility and inherent brittleness of glassy structure (Schuh et al. 2007). This property becomes even more prominent at large length scales (bulk metallic glasses [BMGs]–metallic glasses typically having thickness > 1 mm) (Chen 1974; Drehman et al. 1982; Kui et al. 1984; Wang et al. 2004) as prominent catastrophic failure mechanisms (shear band) dominate (Trexler and Thadhani 2010; Cheng and Ma 2011; Qiao et al. 2016). This severely limits their application toward their use in making large-scale machinery components. This disadvantage can be overpowered by inducing plasticity in glassy structure while retaining its high strength at the same time (Hays et al. 2000; Hofmann et al. 2008a; Hofmann 2010; Wu et al. 2014b). This can be done by various mechanisms including exploitation of intrinsic ability of glass to exhibit plasticity at very small (nano) length scale (Guo et al. 2007; Jang and Greer 2010), introduction of external impulse (obstacles) to shear band formation and propagation (*ex situ* composites) (Choi-Yim 1998; Choi-Yim et al. 2002), self or externally assisted multiplication of shear bands (Lee et al. 2004; Trexler and Thadhani 2010), formation of ductile phases in brittle glassy matrix during solidification (*in situ* composites) (Pauly et al. 2010; Song et al. 2012; Wu et al. 2010, 2016), and transformation inside a ductile crystalline phase, for example, B2–B19′ transformation in Zr-based systems (stress/transformation-induced plasticity) (Kim et al. 2011; Gao et al. 2015; Song et al. 2016; Zhai et al. 2016). The latter approach (formation of ductile phase in brittle glass) takes into account the nucleation of secondary (ductile) phase either during solidification *in situ* (Pekarskaya et al. 2001; Fan et al. 2002; Hu et al. 2003; Zhang et al. 2005; Wu et al. 2007; Zhu et al. 2010; Cheng et al. 2013) or during heat treatment of solidified glassy melt (devitrification/relaxation) (Chen 1976; Antonione et al. 1998; Fan and Inoue 2000; Fan et al. 2000; Basu et al. 2003; Fan et al. 2006b; Gu et al. 2013; Tan et al. 2013; Krämer et al. 2015) and forms the basis of ductile BMG composites.

Although considerable progress has been made in advancing "as cast" sizes in BMG and their composites, the maximum possible diameter and length which has been produced by conventional means till date (Nishiyama et al. 2012) are still far from limit of satisfaction to be used in any structural engineering application. This primarily is associated with mechanical cooling rate achievable as a result of quenching effect from water-cooled walls of Cu container, which in itself is not enough to overcome critical cooling rate (R_c) of alloy (~0.067 K/s [Nishiyama et al. 2012]) to produce a uniform large bulk glassy structure. In addition, occurrence of this bulk glassy structure is limited to compositions with excellent inherent glass-forming ability (GFA) (He, Schwarz, and Archuleta 1996; Inoue et al. 1996). This is not observed in compositions that are strong candidates to be exploited for making large-scale industrial structural components (Inoue et al. 1990; Peker and Johnson 1993; Chen et al. 2009; Tan et al. 2011; Biffi et al. 2013; Cheng and Chen 2013; Wang et al. 2014c; Chu et al. 2015; Jeon et al. 2015; Song et al. 2016) with higher critical cooling rates (R_c) (10 K/s [Peker and Johnson 1993]). This poses a limitation to this conventional technique and urges the need for an advanced manufacturing method that does not encompass these shortcomings. Additive manufacturing has emerged as a potential technique (Gibson et al. 2010a; Spears and Gold 2016) to fulfill this gap and produce BMG matrix composites (BMGMCs) (Schroers 2010; Pauly et al. 2013) in a single step across a range of compositions virtually covering all spectrums (Zheng et al. 2009; Buchbinder et al. 2011; Olakanmi et al. 2015; Li et al. 2016). It achieves this by exploiting very high cooling rates available in a very short period transient liquid melt pool (Li and Gu 2014; Romano et al. 2015; Yap et al. 2015) in a small region where high-energy source laser (LSM/LSF [solid], selective laser melting/laser engineered net shaping [SLM/LENS] [powder] or electron beam melting [EBM]) strikes the sample. This, when coupled with superior GFA of BMGMCs, efficiently overcomes dimensional limitation as virtually any part carrying glassy structure can be fabricated. In addition, incipient pool formation (Romano et al. 2015) and its rapid cooling result in extremely versatile and beneficial properties in final fabricated part such as high strength, hardness, toughness, controlled microstructure, dimensional accuracy, consolidation, and integrity. The mechanism underlying this is layer-by-layer (LBL) formation, which ensures glass formation in each layer during solidification before proceeding to the next layer. This is how a large monolithic glassy structure can be produced. This LBL formation also helps in the development of secondary phases in a multicomponent alloy (Sun and Flores 2010; Yang et al. 2012; Zhang et al. 2015) as layer preceding fusion layer (which is solidified) undergoes another heating cycle (heat treatment) below melting temperature (T_m) somewhat in the nose region of TTT diagram (Pauly et al. 2013), which not only assists in phase transformation (Fan et al. 2006b; Tan et al. 2013) but also helps in increase of toughness, homogenization, and compaction of part. This is a new, promising, and growing technique of rapidly forming metal (Frazier 2014), plastic (Wong and Hernandez 2012), ceramic, or composite (Travitzky et al. 2014) parts by fabricating a near-net shape out of raw materials either by powder method or by wire method (classified on the basis of additives used). The movement of energy source (laser or electron beam) is dictated by a CAD geometry that is fed to a computer at the back-end and maneuvered by CNC (Baufeld et al. 2011; Chen et al. 2011) system. Process has wide range of applicability across various industrial sectors ranging from welding (Li et al. 2006; Kim et al. 2007; Kawahito et al. 2008; Wang et al. 2010b, 2011, 2012a), repair (Acharya and Das 2015; Sames et al. 2016), and cladding (Wu and Hong 2001; Wu et al. 2002; Yue et al. 2007; Zhu et al. 2007b; Yue and Su 2008; Zhang et al. 2011c; Harooni et al. 2016) to full-scale part development.

However, there is dearth of knowledge about exact mechanisms of formation (NG and/or LLT [Wei et al. 2013; Zu 2015; Lan et al. 2016]) of ductile phase dendrites or spheroids *in situ* during solidification of BMGMC happening inside liquid melt pool of additive manufacturing (AM), which is essential for further advance improvement in the process and suggest its optimization. Modeling and simulation techniques, especially those employing finite element methods (phase field [Boettinger et al. 2002; Emmerich 2009; Emmerich et al. 2012; Wang and Napolitano 2012; Gránásy et al. 2014; Gong and Chou 2015]), CAFE [Charbon and Rappaz 1993; Rappaz and Gandin 1993; Gandin and Rappaz 1994, 1997; Gandin et al. 1996; Chen et al. 2014], and their variants at part scale are very helpful in explaining the evolution of microstructure and grain size development in metals and alloys. They have been extensively used in predicting solidification behavior of various types of melts during conventional production methods (Tsai and Hwang 2011; Chen et al. 2014; Chen 2014; Wei et al.). However, their use in AM applications (Zhang et al. 2013a; Wang et al. 2014d; Zhou et al. 2016; Zinoviev et al. 2016) specially related to BMGMC is still in its infancy. Virtually no effort has been made to understand nucleation and growth of ductile crystalline phase dendrites or spheroids *in situ* during solidification in BMGMC by modeling and simulation. A step forward is taken in the present study to address these gaps and bring together the strengths of different techniques and methodologies at one platform. An effort is made to form ductile BMG metal matrix composites by taking advantage of

a. Materials Chemistry: A Multicomponent Alloy. Its GFA is used as a measure to manipulate composition and vice versa.
b. Solidification Processing: Liquid melt pool formation, its size, shape and geometry, role of number density, size and distribution of ductile phase in resultant glassy alloy matrix. It is taken as a function of type, size, and amount of nucleates (inoculant).
c. Additive Manufacturing: Use of very high cooling rate inherently available in the process to (a) not only form glassy matrix but also use liquid melt pool formed at very high temperature to trigger nucleation (liquid–solid transformation) of ductile phase in the form of dendrites or spheroids from within the pool "*in situ*" (This is done by controlling machine parameters in such a way that optimized cooling rate satisfying narrow window of "quenching" BMGs is achieved.); (b) take advantage of heating (heat treatment) of preceding layer to trigger solid–solid transformation (devitrification) again to form ductile phase and achieve homogeneity, consolidation, and part integrity, eliminating the need of postprocessing or after treatment.
d. Modeling and Simulation: Strong and powerful mathematical modeling techniques based on
   a. Transient heat transfer for "liquid melt pool formation as a result of laser–matter interaction"
   b. Its "evolution–solidification" by
      i. Deterministic (modified CNT, JMAK correction, and Rappaz modification) or
      ii. Stochastic/probabilistic (3D CAFE model for nucleation and growth [solute diffusion and capillary action driven]) modeling of microstructure evolution and grain size determination of ductile phase equiaxed dendrites or spheroids in glassy melt will be used to simulate the conditions in liquid melt pool of BMGMC during AM. Effect of number density, size, and distribution of ductile phase dendrites or spheroids will be evaluated/verified using simulation of melt pools developed using different values of aforementioned parameters.

Note:

a. AM methods can also be classified on the basis of energy source used (i.e., laser based or electron beam based).

b. Heating in nose region of TTT diagram can also trigger assimilation of free volume (relaxation) whose effect is not taken into account in the present study due to difference in mechanism in which it occurs. It does not constitute any chemical reaction.

BULK METALLIC GLASSES AND BULK METALLIC GLASS MATRIX COMPOSITES

1.1 METALLIC GLASSES AND BULK METALLIC GLASSES/MONOLITHS

Metallic glasses (MG) (Chen 1974) may be defined as "disordered atomic-scale structural arrangement of atoms formed as a result of rapid cooling of complex alloy systems directly from their melt state to below room temperature with a large undercooling and a suppressed kinetics in such a way that the supercooled state is retained/frozen" (Güntherodt 1977; Greer 1995; Inoue 1995; Johnson 1999). This results in the formation of "glassy structure." The process is very much similar to inorganic/oxide glass formation in which large oxide molecules (silicates/borides/ aluminates/sulfides and sulfates) form a regular network retained in its frozen/supercooled liquid state (Matthieu 2016). The only difference is that MGs are comprised of metallic atoms rather than inorganic metallic compounds. Their atomic arrangement is based on a mismatch of atomic size and quantity (minimally three) (Hofmann and Johnson 2010) (described in the next section), is based on short-range order (Shi and Falk 2006; Mattern et al. 2009; Jiang and Dai 2010) to medium-range order (Sheng et al. 2006; Cheng et al. 2009; Zhang et al. 2014a) or long-range dis-order (Inoue and Takeuchi 2011) (unlike metals—well-defined long-range order), and can be explained by other advanced theories/mechanisms (frustration [Nelson 1983], order in disorder [Nelson 1983; Sheng et al. 2006; Ma 2015], and confusion [Greer 1993]). Important features characterizing them are their amorphous microstructure and unique mechanical properties. Owing to the absence of dislocations, no plasticity is exhibited by bulk metallic glasses (BMGs). This results in very high yield strength and elastic strain limits as there is no plane for material to flow (by conventional deformation mechanisms). From a fundamental definition point of view, MGs are typically different from BMG in that the former has fully glassy (monolithic) structure for thicknesses less than 1 mm, while the latter is glassy (monolithic) in greater than 1 mm (Drehman et al. 1982; Kui et al. 1984). To date the largest BMG made in "as cast" condition is 80 mm in diameter and 85 mm in length (Nishiyama et al. 2012). There are reports of making large thin castings as casing of smart phone but they are typically less than 1 mm (Qiao et al. 2016). Furthermore, they are characterized by special properties such as glass-forming ability (GFA) and metastability (which will be described in following sections).

1.2 THREE LAWS

The formation and stability of BMG (even in metastable condition) is described by their ability to retain glassy state at room temperature. Although the understanding of glass and glassy structure was established much earlier, it was very difficult to form homogeneous, uniform glassy structure across whole section thickness at room temperature until recently. Only alloys of very narrow compositional window cooled at extremely high cooling rate can form glassy structure (Klement et al. 1960; Turnbull 1969; Chen 1974, 1980; Drehman et al. 1982). Any deviation from any of these parameters severely hampers the retention of glassy state and crystallization occurs (Akhtar et al. 1982a, 1982b; Akhtar and Misra 1985). This property is known as GFA (Inoue, Zhang, and Masumoto 1993). This is the single most important property in MG family of alloys which governs their formation and evolution. GFA has been increasingly studied and considerable progress has been made in its improvement (Donald and Davies 1978; Lu et al. 2008; Wang et al. 2012b; Yi et al. 2016b) by alterations in both composition and window of processing condition (Park and Kim 2005; Chen 2011; Inoue and Takeuchi 2011). Now, alloys having multicomponent composition can be cast in glassy state even at slow cooling rate owing to their superior GFA (Peker and Johnson 1993; Yi et al. 2016; Jia et al. 2006a; Cheng et al. 2008; Park et al. 2008; Miracle et al. 2010; Guo et al. 2016), which in turn is governed by various theories (Donald and Davies 1978; Li et al. 1997, 2011, 2012, 2014a; Fan et al. 2001; Lu and Liu 2002, 2004; Kim et al. 2003; Xu et al. 2004; Park et al. 2008; Yang et al. 2010; Wu et al. 2014a; Shen et al. 2015) and analytical models (Zhang et al. 2013b; Amokrane et al. 2015).

Fundamentally, research over a period of time has yielded three basic laws that are now considered universal for the formation of any BMG system (Hofmann and Johnson 2010). These are described below. Any glass-forming system consists of elements that must

1. Be three in number (at minimum). (Elements greater than 3 are considered beneficial.)
2. Differ in their atomic size by 12% among three elements. (Atoms of elements with large size are considered to exhibit superior GFA.)
3. Have negative heat of mixing among three elements. (This ensures tendency to demix or confuse (Greer 1993) ensuring retention of glassy structure at room temperature.)

This results in new structure with high degree of densely packed atomic configurations, which, in turn, results in completely new atomic configuration at the local level with long-range homogeneity and attractive interaction. In general BMG or bulk glassy alloy (BGA) is typically designed around (1) alloy systems that exhibit a deep eutectic, which decreases the amount of undercooling needed to vitrify the liquid, and (2) alloys that exhibit a large atomic size mismatch, which creates lattice stresses that frustrate crystallization (Hofmann and Johnson 2010). An important way to arrive at optimum glass-forming composition and then selecting alloying elements is based on proper choice of eutectic or off-eutectic composition, diameter, and heat of mixing (Inoue and Takeuchi 2011). These laws were first proposed by Douglas C. Hoffmann and his supervisor at Caltech (Hofmann and Johnson 2010) and Prof. Akisha Inoue at WPI IMR, Tohoku University, Japan (Inoue and Takeuchi 2011) independently.

1.3 CLASSIFICATION

As proposed by Prof. Inoue (Inoue and Takeuchi 2002, 2011; Inoue et al. 2006), BMG can be broadly **classified** into **three** types (Figure 1.1):

1. Metal–Metal Type
2. Pd–Metal–Metalloid Type
3. Metal–Metalloid Type

This classification is based on the ease with which one group of metals reacts with the other group to finally evolve a glassy structure, which in turn is chosen by various rules such as chemical affinity, atomic size, electronic configuration, etc. Their proposed atomic arrangement, size, and crystal structure are shown in Figure 1.1. *Metal–metal type*

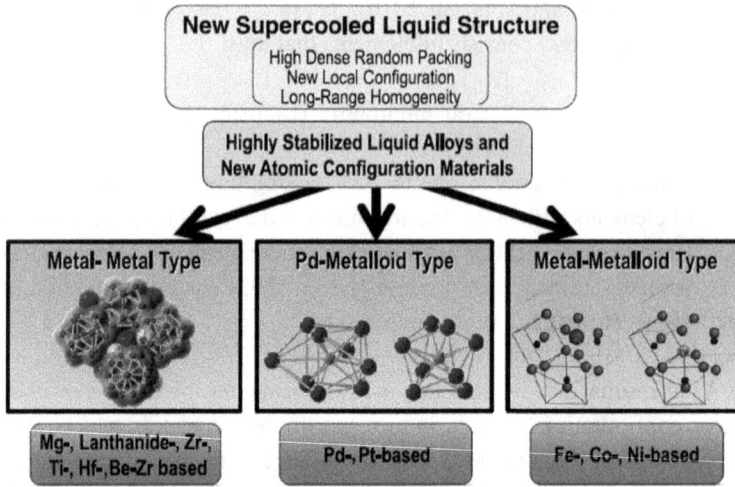

Figure 1.1. Classification of BGAs (Inoue and Takeuchi 2002, 2011).

glassy alloys are composed of icosahedral-like ordered atomic configurations. They are exemplified by Zr–Cu–Al–Ni and Zr–Cu–Ti–Ni–Be type systems. *Pd–transition metal–metalloid* type glassy alloys consist of high dense packed configurations of two types of polyhedra of Pd–Cu–P and Pd–Ni–P atomic pairs. Their typical examples are Pd–Cu–Ni–P systems. *Metal–metalloid type* glassy alloys have network-like atomic configurations in which a disordered trigonal prism and an anti-Archimedean prism of Fe and B are connected with each other in face- and edge-shared configuration modes through glue atoms of Ln and early transition metals (ETM) of Zr, Hf, and Nb. Their typical examples are Fe–Ln–B and Fe–(Zr, Hf, Nb)–B ternary systems. These icosahedral-, polyhedral-, and network-like ordered atomic configurations can effectively suppress the long-range rearrangements of the constituent elements that are necessary for onset of crystallization process. Among the three structures described, the second and third types have similarity that they both contain trigonal prism structure but are different in that the latter forms a well-developed connected structure of prisms by sharing their vertices and edges, which results in highly stabilized supercooled liquid leading to the formation of BGA even at very slow cooling solidification processes (Inoue and Takeuchi 2011). From an **engineering** standpoint, BGA adopts another system of classification that is based on their applicability. They are classified into seven types which in turn are grouped into two main types based on their behavior in phase diagrams. These are described as follows:

a. Host metal base type: Zr–Cu–Al–Ni, Fe–Cr–metalloid, Fe–Nb–
   metalloid, and Fe–Ni–Cr–Mo–metalloid systems and

b. Pseudo-host metal base type: Zr–Cu–Ti–Ni–Be, Zr–Cu–Ti–(Nb,
   Pd)–Sn, and Cu–Zr–Al–Ag systems

It can be observed that Fe and Zr comprise the most important
materials for practical use. Further subclassification of Zr-based BMG is
also proposed by Prof. Inoue whose details can be found in cited literature
(Inoue and Takeuchi 2011).

1.4 IMPORTANT CHARACTERISTICS

The formation and stability of BMGs are governed by their ability to form
complex network and then retain this below room temperature. This is best
described by its intrinsic properties specific to these alloy systems. These
are GFA and metastability.

1.4.1 GLASS-FORMING ABILITY

As described in Section 1.2, GFA may be defined as "inherent, intrinsic
ability of a multicomponent system to consolidate in state of low energy
in such a way that glass formation is promoted and crystallisation is
retarded." This single unique parameter is effectively used to identify and
design a range of glassy alloys. The GFA of a melt is evaluated in terms
of the critical cooling rate (R_c) for glass formation, which is *the minimum
cooling rate necessary to keep a constant volume of melt amorphous
without precipitation of any crystals during solidification* (Weinberg
et al. 1989, 1990; Kim et al. 2005a; Ray et al. 2005; Zhu et al. 2007a). In
addition, they must possess inherent resistance against crystallization, that
is, their atomic configuration should be such that they should not favor its
rearrangement in regular crystallographic patterns. GFA is a strong function of another parameter known as "overall cooling power" or "strength
of quench." Generally,

$$\text{GFA} \propto \text{strength of quench} \qquad (1)$$

which means, the higher the quenching power, better will be the ability of a
material to form glass. However, this is not a hard and fast rule and excep-
tions exist (Peker and Johnson 1993; Jia et al. 2006a; Cheng et al. 2008;

Park et al. 2008; Miracle et al. 2010; Guo et al. 2016; Yi et al. 2016) (as described in Section 1.2). For example, in a well-defined multicomponent system, for example, Zr–Ti–Cu–Ni–Be (Peker and Johnson 1993), BMG can be formed even at slower cooling rate. This is because, the above-mentioned two criteria are effectively fulfilled in these BMGs, while in others, for example, Ti- and Cu-based BMG, glassy structure can only form in relatively thin sections (because of very high cooling rates experienced there)—which is essential for glass formation in these systems. Also, these systems do not exactly meet above criteria and deviations exist, which promote their inability to form glassy structure even upon fast cooling. Metals that most commonly account for formation of BMG are ETM and late transition metals (Inoue 2000; Inoue and Takeuchi 2002; Inoue et al. 2006). From a phase development point of view, they always include a eutectic point with the lowest liquidus temperature. An important factor to design these alloys is to choose a composition exhibiting a lower liquidus temperature in the vicinity of eutectic point. Although variants exist (off-eutectic compositions) (Ma et al. 2003, 2011; Wang et al. 2004; Jian 2009; Lee et al. 2012), this method is effective to an appreciable extent for the design of BGA/BMGs (Inoue and Takeuchi 2011).

There have been different theories in the way GFA has been predicted over the years. For example, David Turnbull in his classical paper (Turnbull 1969) mentioned the use of reduced glass transition temperature (T_{rg}) where it is defined as the ratio of glass transition temperature (T_g) and liquidus temperature (T_l):

$$T_{rg} = \frac{T_g}{T_l} \tag{2}$$

This still has been the basic method of determining GFA to a large extent. However, there have been limitations around it and there were other theories that were predicted. For example, the use of *supercooled liquid region* ΔT_x (the temperature difference between the onset of crystallization temperature T_x and glass transition temperature T_g) (Inoue 2000).

$$\Delta T_x = T_x - T_g \tag{3}$$

The γ parameter (Lu and Liu 2002) is defined as

$$\gamma = \frac{T_x}{(T_g + T_l)} \tag{4}$$

None of these alone or combined is good enough to predict the GFA of BMGs (Donald and Davies 1978; Inoue et al. 1993; Lu and Liu 2003; Lu et al. 2008) and the GFA remains a function of alloy composition to a large extent empirically, which keeps on changing (Cheng et al. 2008; Wang et al. 2012b; Lad 2014; Li et al. 2014a; Wu et al. 2014a; Zhang et al. 2014a; Guo et al. 2016). Following diagrams can be effectively used to arrive at the nearest possible composition at which BMG alloy formation is expected in the mentioned ternary (Figure 1.2) and quaternary systems (Figure 1.3).

From a phase transformation point of view, they follow ternary phase diagrams more predominantly than binary diagrams because of constraints posed by necessity of having three elements. Their mechanical properties can also be explained on the basis of ternary phase diagrams more effectively. When used in conjunction with the compositional contrast diagrams (Figures 1.2 and 1.3), these can effectively predict a suitable alloy system that will show superior GFA along with a set of mechanical properties (Inoue and Takeuchi 2011).

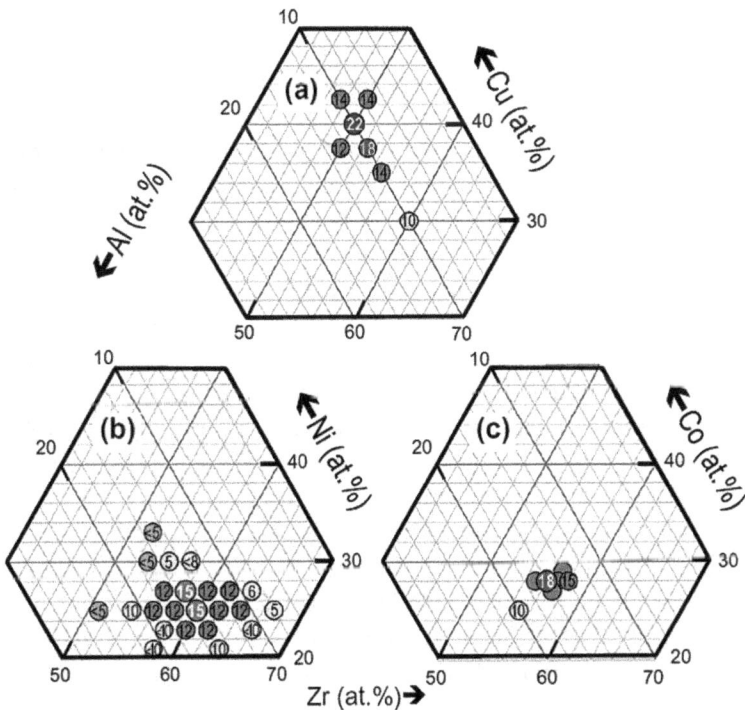

Figure 1.2. Composition range in which the BGAs are formed by the copper mold casting method and the composition range of the maximum diameter of cast glassy alloy rods in Zr–Al–Cu, Zr–Al–Ni, and Zr–Al–Co systems (Inoue and Takeuchi 2011).

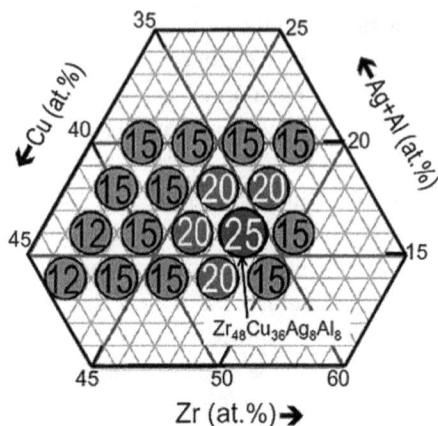

Figure 1.3. Compositional dependence of maximum diameter of Zr–Cu–Al–Ag glassy alloys produced by copper mold casting (Inoue and Takeuchi 2011).

1.4.2 METASTABILITY

Another important characteristic of these glass-forming systems is their composition which also describes their inhomogeneity and metastability. They are not cooled to room temperature following equilibrium phase diagram, but their formation and evolution are governed by nonequilibrium diagrams also known as time temperature transformation (TTT) diagrams (Mukherjee et al. 2004). This gives rise to metastable structures resulting from very high cooling rates (Baricco et al. 2004; Brazhkin 2006a, 2006b). These metastable structures are the reason for the extremely high strength of these systems. Upon heating, BMGs relax their structural disorder/misfit and give rise to ordered structures. This process is known as devitrification (Taub and Spaepen 1980; Tsao and Spaepen 1985; Akhtar and Misra 1986; Wen et al. 2010; Qiao and Pelletier 2014; Liu et al. 2015). This also helps further to explain and understand the development and formation of quasicrystals (QC) (Levine and Steinhardt 1985; Steinhardt 1990; Janot 1993; Xing et al. 1998, 1999; Fan et al. 2001) and ductile phases (e.g., B2 CuZr, β-Zr) that are responsible for increase in ductility and toughness of BMGMC. It is also very important in defining the behavior of BMGMC in additive manu-facturing as material undergoes repeated thermal cycles that vitrify the liquid and devitrify the glass.

1.5 LIMITATIONS

Despite their advantages and extremely high strength, MG and their bulk counterparts suffer from the following limitations:

a. They have very poor ductility (Xi et al. 2005; Gu et al. 2007; Schuh et al. 2007; Guo et al. 2014). They do not exhibit any plasticity under tension and exhibit little plastic behavior under compression (Chen 1973; Xue et al. 2008; Chen and Tsai 2015).
b. They have very poor fracture toughness (Kimura and Masumoto 1975; Conner et al. 1997; Gilbert et al. 1997; Lowhaphandu and Lewandowski 1998; Lowhaphandu et al. 2000; Flores and Dauskardt 2004; Hofmann et al. 2008a; Xu et al. 2010). This severely limits their engineering applications as they cannot absorb effects of load or cannot transfer stresses safely and fail in a catastrophic manner (Chen 2008).

Progress has been made during recent years to overcome these problems, but still experimental results and values obtained so far are not of considerable practical significance and have poor reproducibility, which render them unsatisfactory for any practical use (Ritchie 2011; Sarac and Schroers 2013; Hufnagel et al. 2016).

1.6 DUCTILE BMGs

Owing to difficulties encountered during the use of "as cast" BMGs especially for structural applications, schemes were devised from very early days of BMG research for the increase of ductility in these alloys. In the beginning, efforts were made to increase the plasticity by dispersing controlled porosity (Wada et al. 2005), but these efforts did not proceed far because of nonpractical nature of method and other unwanted problems that developed in the structure. Then, the focus was directed to address this problem by basic mechanisms of plasticity and plastic deformation. For example, if progression of shear band could be hindered (just like dislocation motion hindrance in crystalline alloys) by impeding its motion, a substantial increase in ductility could be achieved. This is achieved by two fundamental mechanisms. (a) Increased number of shear bands increases the obstacles ("arrests") to the paths of material flow. Hence, it would be difficult for the material to flow (Spaepen 1977; Donovan and Stobbs 1981, 1983; Steif et al. 1982; Leng and Courtney 1991; Huang et al.

2002; Liu et al. 2005; Flores 2006); (b) strain energy dissipation resulting from shear band formation at the interface between crystalline phase and amorphous matrix. This gives rise to new processes of shaping/forming by controlled application of force (thermoplastic forming [TPF]) (Sarac et al. 2011; Li et al. 2016). This new method of fabricating BMGs is also known as super plastic forming (Schroers 2005) which was tried as far as 10 years ago. Further techniques consisted of (1) *ex situ* introduction of second-phase reinforcements (particles [Dandliker, Conner, and Johnson 1998; Choi-Yim et al. 2002; Jiang and Qiu 2015], flakes [Sun et al. 2006], fibers [Conner et al. 1998; Wadhwa et al. 2007; Lee et al. 2010], ribbons [Cytron 1982], whiskers [Deng et al. 2011; Wang et al. 2016d], etc.), which offer a barrier to movement of shear bands along one plane and provide a pivot for their multiplication; (2) *in situ* nucleation and growth of second-phase reinforcements in the form of equiaxed dendrites which are ductile in nature; thus, they not only provide means of increase in ductility by themselves but offer a pivot for multiplication of shear bands (explained in the next section) (Qiao 2013; Jiang et al. 2015); (3) reducing the size of glass to nanometer and ductile phase to micrometer (Zhai et al. 2016); (4) making the plastic front (local plastically deformed region ahead and around shear band) of shear bands to match with difficult plane of flow in crystal structure of ductile phase, thus creating an easy path for the shear band to multiply (not yet investigated idea of author), and (5) heating the alloy to cause temperature-induced structural relaxation (devitrification) (Freed and Vander Sande 1979; Taub and Spaepen 1980; Akhtar and Misra 1986; Qiao and Pelletier 2014; Liu et al. 2015), etc. The drive for all these mechanisms is different. For example, it is natural that shear bands are responsible for the catastrophic failure of BMGs (Greer et al. 2013) and any hindrance to their motion or simply increase in their number all spread across the volume of material would cause a difficulty with which they move (along one direction at very high speed), causing abrupt failure. This gives rise to fundamental mechanisms of toughening (Hofmann et al. 2008a; Gludovatz et al. 2014). Similar effect could be achieved by external addition to (*ex situ*) or internal manipulation of (*in situ*) structure of material. Of these, only structural relaxation was first envisaged as dominant mechanism for increase of ductility. It was known thermodynamically and observed numerically (Ogata et al. 2006) and experimentally (Pampillo 1972; Zhou et al. 1998; Packard and Schuh 2007; Pan et al. 2008) since early days that structurally constrained glass relaxes during heating known as "devitrification." The drive for devitrification (Taub and Spaepen 1980; Tsao and Spaepen 1985) did not came out of ingenious effort but was natural impulse as BMG possesses natural tendency to undergo release of stresses with formation of new structures (phases) (solid state phase transformations) when subjected to temperature effect similar to heat treatment for crystalline metallic alloys. This results in a new class of BMG called **ductile BMG** (Schroers and Johnson 2004; Das et al. 2005b, 2007; Jiang et al. 2006; Yao et al. 2006; Chen et al. 2008; Abdeljawad et al. 2011; Lu et al. 2013; Magagnosc et al. 2013). The research on other mechanisms was adopted, or is envisaged to be investigated with passage of time giving rise to more versatile materials known as ductile BMG composites (explained in the following).

1.7 DUCTILE BULK METALLIC GLASS MATRIX COMPOSITES

As minutely introduced in the previous section, a significant improvement in mechanical properties of BMGs was reported first in 2000 by Prof. William's Group at Caltech (Hays et al. 2000) when they successfully incorporated ductile second-phase reinforcements in glassy matrix in the form of precipitates *in situ* nucleating during solidification, thus giving birth to "so-called" family of *in situ* dendrite/MG matrix composites. These materials are formed as a result of conventional solute partitioning mechanisms as observed in other metallurgical alloys, resulting in copious formation of ductile phase (Ti–Zr–Nb β in case of Ti-based composites (Hays et al. 2000), B2 Cu–Zr in case of Zr-based composites (Jiang et al. 2007; Liu et al. 2010a, 2012a, 2014; Song et al. 2011; Ding et al. 2014), or transformed B2 (B19′ martensite) in case of Zr–Cu–Al–Co shape memory BMGMC (a special class of BMGMC) (Schryvers et al. 1997; Seo and Schryvers 1998a, 1998b; Liu et al. 2010a; Pauly et al. 2010; Song 2013; Song et al. 2013) predominantly (not always) in the form of three-dimensional (3D) dendrites or spheroids emerging directly from liquid during solidification. Devitrification in these alloys can be explained by "phase separation" before solid state transformation or "quenched in" nuclei (Kelton 1998; Park et al. 2006; Antonowicz et al. 2008; Kim et al. 2013; Sun et al. 2016). This is another very important route for the fabrication of these alloys. They also comprise of the family of BMG composites that are formed by more advanced transformation mechanisms (liquid state phase separation) (Kündig et al. 2004; Oh et al. 2005; Park and Kim 2006; Chang et al. 2010) which has recently become observable owing to more advanced characterization techniques using synchrotron radiation (Mu et al. 2013; Michalik et al. 2014; Guo et al. 2015a, 2015b) and container less levitated sample solidification in micro and zero gravity conditions (Paradis et al. 2014; Zu 2015). These render them special properties (enhanced plasticity and compressive strength) not otherwise attainable by other conventional processing routes or in simple binary and ternary compositions. This, however, is seldom the case and is not readily observed as compared to solid state phase separation (Chang et al. 2010), which is the dominant mechanism in these alloys. More advanced mechanisms of forming these materials are by local microstructural evolution by phase separation right at shear bands (Yi et al. 2016). It narrates that solid–solid phase separation (spinodal decomposition) occurs at the onset of shear band, which is the cause of microstructural evolution.

Few notable class of alloys in these types of ductile composites are Ti-based BMGMCs (He et al. 2003; Kim et al. 2005b; Huang et al. 2007; Oh et al. 2011; Wang et al. 2014b, 2014c; Chu et al. 2015; Zhang et al. 2014b), Ti-based shape memory BMGMC (Gargarella et al. 2013), Zr–Cu–Al–Ti (Hofmann et al. 2008b; Chu 2009), Zr–Cu–Al–Ni (Cheng and Chen 2013), and Zr–Cu–Al–Co shape memory BMG composites (Biffi, Figini Albisetti, and Tuissi 2013). Each has its own mechanisms of formation and individual phases are formed by liquid–solid (L–S) or solid–solid (S–S) phase transformations.

1.8 PRODUCTION METHODS—MECHANISMS PERSPECTIVE

From mechanisms perspective, production and processing of BMG matrix composites (BMGMCs) can be characterized into two fundamental types:

1.8.1 LIQUID–SOLID TRANSFORMATION (SOLIDIFICATION)

This mechanism is typically characterized by solidification of melt from liquid state to room temperature. The resulting metal–matrix composite may consist of melt which carry solid particles suspended, or homogeneously mixed in the melt prior to or during casting as a result of very chemical nature of alloy (limit of solid solubility enabling formation of substitutional or interstitial solid solution upon solidification). These are called "*ex situ*" and "*in situ*" BMGMC respectively.

1.8.1.1 Ex Situ Formation

These are methods of production in which external particles (reinforcements) in various forms (particles, flakes, fibers, and whiskers [Section 1.6]) are introduced in the bulk melt and are homogenized by various means before final shape casting. These are versatile in a sense that properties could be manipulated by controlling type, size, shape, and amount of external reinforcements. Furthermore, various types of numerical and/ or statistical means (permutations and combinations) could be applied to control the material processing and fabrication. Melt infiltration is one of the popular routes for the fabrication of these composites. This is one of the well-established techniques in ceramics processing (Kingery 1960). This was one of the first methods adopted again at Caltech (Dandliker et al. 1998) in 1997–1998 to fabricate their world famous Vitreloy 1 ($Zr_{41.2}$ $Ti_{13.8}Cu_{12.5}Ni_{10.0}Be_{22.5}$) as matrix with continuous ceramic (SiC and carbon), metal (tungsten [254 μm Ø], carbon steel [AISI 1080] [254 μm Ø] [in most of the cases], stainless steel, molybdenum, tantalum, nickel, copper, and titanium) wires cut into 5 cm length as well as with loose tungsten (W) powders and sintered silicon carbide (SiC) particulate preforms as reinforcements. They infiltrated different specimens with aforementioned reinforcements and casted rods in size 5 cm in length and 7 mm in diameter. The process was carried out in vacuum induction furnace with titanium gettered Ar atmosphere. The starting metals were high-purity (≥99.5% metal basis) research-grade material. The composite fabricated were quenched after infiltration. A schematic of setup used for this purpose is shown in Figure 1.4.

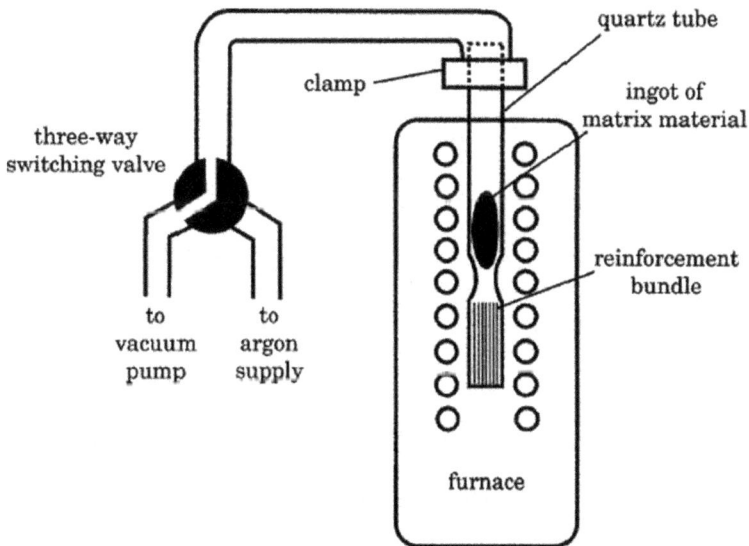

Figure 1.4. Schematic of setup for melt Infiltration (Dandliker, Conner, and Johnson 1998).

The resulting composite was analyzed with X-ray diffraction (XRD) and scanning electron microscopy (SEM). The measured porosity was ≤3% and matrix was about 97% amorphous material. The detail of procedure could be found in Dandliker et al. (1998). They also carried out the mechanical (compressive and tensile) and structural (SEM) characterization and fractography of the composite developed in another study (Conner et al. 1998). It was found that tungsten reinforcement increased compressive strain to failure by over 900% compared to the unreinforced monolithic Vitreloy 1. A definite increasing near-linear relationship was observed between fiber volume fraction (V_f) and elastic modulus (E) (Figure 1.5) and yield stress (Figure 1.6). The increase in compressive toughness was attributed to fibers restricting propagation of shear bands, which in turn promotes generation of further multiple shear bands and additional fracture surface area. There is direct evidence of viscous flow of the MG matrix inside regions of the shear bands. Samples reinforced with steel were found to have an increased tensile strain to failure and energy to break by 13% and 18%, respectively. Reason for the increase in tensile toughness was found to be ductile fiber delamination, fracture, and fiber pullout as observed by fractography under SEM (Figure 1.7).

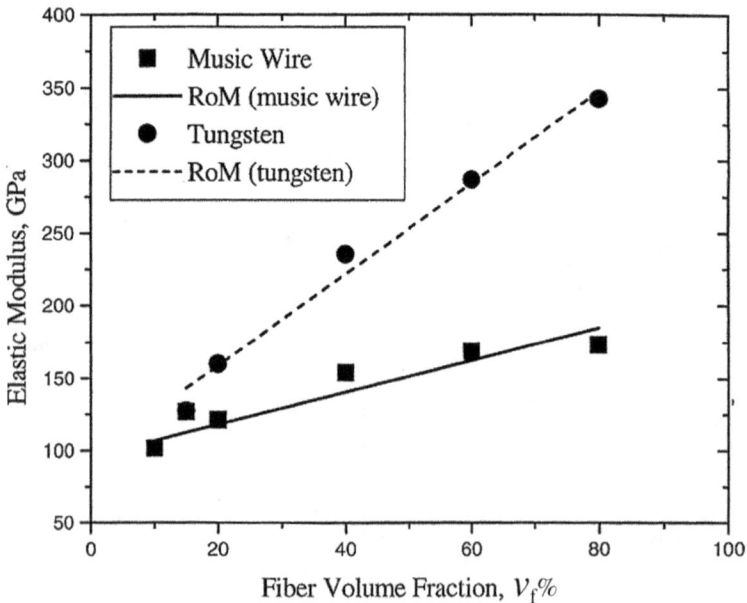

Figure 1.5. Relationship between volume fraction (V_f) of fiber (percentage) and elastic modulus expressed in GPa. Dashed lines are rule of mixture (RoM) modulus calculations (Conner et al. 1998).

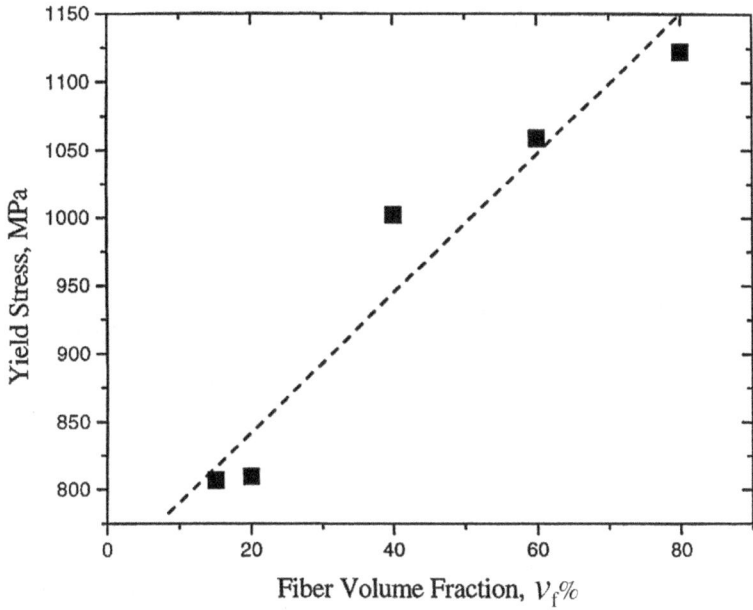

Figure 1.6. Relationship between volume fraction (V_f) of fiber (percentage) and yield stress (MPa) (Conner et al. 1998).

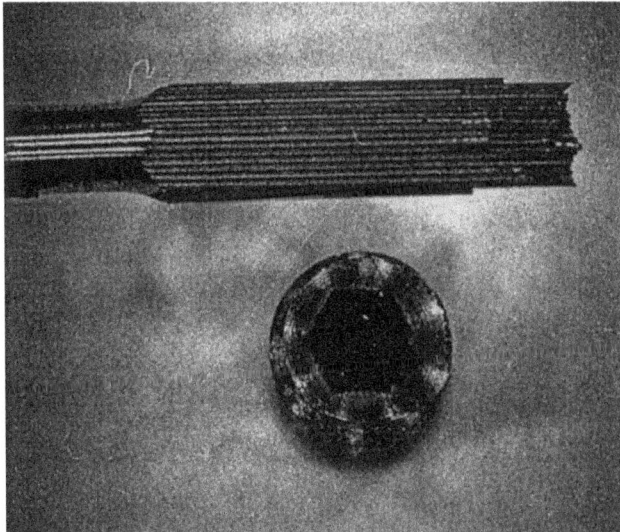

Figure 1.7. Fiber pullout of 80% steel wire/Vitreloy composite tensile specimen (Conner et al. 1998).

In a further study (Conner et al. 1999), they modified the alloy composition to $Zr_{57}Nb_5Al_{10}Cu_{15.4}Ni_{12.6}$ and repeated the same procedure of melt infiltration without AISI 1080 carbon steel wire but with Ta particles. Other reinforcements (SiC and W) remained the same in varying volume fractions. The samples were tested under compression and tension as done previously. It was found that compressive strain to failure increased by more than 300% compared with the unreinforced $Zr_{57}Nb_5Al_{10}Cu_{15.4}Ni_{12.6}$ monolithic BMG and energy to break the tensile samples increased by more than 50%. A typical fracture surface of $W/Zr_{57}Nb_5Al_{10}Cu_{15.4}Ni_{12.6}$ specimen is shown in Figure 1.8, indicating direction of shear band propagation.

This study was extended in 2001 (Choi-Yim et al. 2001) by another postdoc Ms. Haein Choi-Yim and she performed quasistatic as well as dynamic tests. The aim was to manufacture "so-called" kinetic energy penetrators (KEP) under a program funded by the US Army Research Office (under STIR program) and German Science Foundation (DFG). BMGs were selected as suitable material owing to their very high hardness. The material was changed from previously reported reinforcements to 80% V_f W wire (254 μm Ø), 50% V_f W particles (avg. Ø 100 μm), and 50 V_f mixed W/Rhenium (avg. Ø 50 μm (W) and 10 μm (Re) particles while matrix remained the same ($Zr_{57}Nb_5Al_{10}Cu_{15.4}Ni_{12.6}$ (Vit106). Dynamic testing was performed in the form of ballistic tests by firing BMG KEPs onto Al 6061 T651 and 4130 steel targets. Results indicated a new "self-sharpening" behavior due to localized shear band failure. Particulate

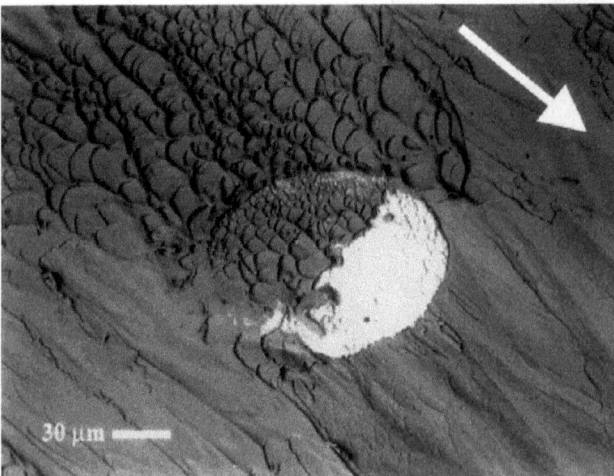

Figure 1.8. Fracture surface of W/Vitreloy 1 showing direction of shear band propagation (Conner et al. 1999).

composite consisting of Zr BMG/50 V_f (W/Re) particles exhibited 8% plastic elongation when tested in quasistatic compression. Strain to failure of 80% V_f W wire/Zr BMG composite with wire aligned with loading axis exceeded 16%. It was more than six times higher than for monolithic BMG alone. Young's Modulus of W wire composite was four times greater than unreinforced matrix material and was in good agreement with RoM prediction. The penetration performance of KEP was 10–20% better than that of W heavy alloy penetrator alone. The study was further continued in 2002 (Choi-Yim et al. 2002) and Vit106 was reinforced with 50 V_f of Mo, Nb, and Ta particles. A detailed characterization based on XRD, differential scanning calorimetry (DSC), electron probe microscopic analysis (EPMA) and scanning electron microscopy (SEM) was performed. The composites were tested in compression and tension. Compressive strain-to-failure increased up to a factor of 12 compared to unreinforced Vit106 BMG. This was due to restriction offered by particles to shear band propagation and generation of further shear band as a consequence. The detail of the study and experimentation could be found in Choi-Yim et al. (2002). Based on these studies, further investigations in the field of ex situ BMGMC picked up interest of various research groups in the world. For example, a research based on studying the relationship between flowing velocities and processing parameters such as pressure, V_f of fibers, and infiltration length was carried out at UST, Beijing in 2006 (Hui et al. 2006). The matrix used was $Zr_{47}Ti_{13}Cu_{11}Ni_{10}Be_{16}Nb_3$ (Vitreloy 1) and reinforcements were W particles. Experimental measurements of contact angle and surface tension of BMG on W were conducted using sessile drop technique at temperature of 1,023 and 1,323 K with different processing times. Wetting angle between melt and W was attributed to belong to reactive wetting as a result of observation of diffusion band at the fringe of metallic drop. This was first good contribution toward studying the chemistry (surface tension and wetting angle) behavior of BMGMCs. This triggered collaborative work between the United States and China on these classes of materials and various studies aimed at different processing and structural parameters of BMGs such as effect of hydrostatic extrusion on porous W/Zr BMG melt-infiltrated composites (resulting in much improved fracture strength [1,852 MPa] due to stable interface between MG and W phase and high dislocation density of W phase) (Xue et al. 2007) and effect of temperature on quasistatic and dynamic testing (Hou et al. 2007) of W/BMGMCs were conducted. Some modeling studies were also conducted by notable research groups (Zhang and Greer 2007) aiming at predicting the selection of composite based on plasticity rather than brittleness. Their details could be found in citations of original work in 2006. During the same time, Prof. Schuh at MIT in collaboration with their Singapore–MIT alliance and NUS, Singapore conducted a detailed dynamic study of different dispersion *in situ* added to Zr-based BMG alloys (Fu et al. 2007). They studied the effect of strain rates, temperature, and different levels of volume fraction (V_f) of Al (precipitated as Al intermetallics [IMC]) as reinforcements in $Zr_{49}Cu_{(51-x)}Al_x$ ($x = 6, 8, 10$, or 12 at%) alloys. All experiments exhibited that mechanical properties are dominated by deformation of the amorphous matrix phase. This includes inhomogeneous flow and fracture at low temperatures and homogeneous flow of both Newtonian and non-Newtonian nature at high temperatures. This strengthening effect (in homogeneous deformation range) caused by Al particles increases as V_f increases and was quantitatively explained in both the Newtonian and non-Newtonian regimes. A shear transformation zone model developed previously by Prof. Ali S. Argon (Argon 1979) was also used to describe the behavior and was coupled with model for reinforced particles. This was found to arise from: (i) load transfer from the amorphous matrix to the reinforcements; and (ii) a shift in the glass structure and properties upon precipitation of the reinforcements. An additional mechanism termed as "*in situ* precipitation" during deformation was held responsible for the increase of strength. This was the first contribution of its kind to the field of BMGMCs by MIT, its alliance and collaborators under US Army Research Office. Research is still going on for this class of material and few notable studies in recent time on this topic have been reported (Liu et al. 2009; Zhang et al. 2009; Khademian and Gholamipour 2010; Chen et al. 2011, 2012). Much attention has been focused on studying their "so-called" *in situ* behavior as exemplified by surface tension, surface chemistry, hydrophilic and hydrophobic response, and wetting angle studies (Liu et al. 2009; Zhang et al. 2009) as well as increase of "plasticity" as predicted by both experimental (static and dynamic tensile and compressive) and simulation studies (Khademian and Gholamipour 2010; Lee et al. 2010; Zhu et al. 2010; Deng et al. 2011).

1.8.1.2 In Situ *Formation*

Research from *ex situ* reinforced composites shifted to so-called *in situ* reinforced composite due to increased plasticity observed in this class of materials. The impulse for this was served not only by the mechanism (chemical reaction) but also by the ease with which these materials are fabricated coupled with observation of enhanced properties (plasticity/ductility) by single step process (i.e., nucleation and growth of dendrites directly from melt during solidification) rather than external mixing (mechanical or physical [infiltration]) of reinforcements which not only is difficult but is an energy-extensive process adding an additional cost to process. The interest was also triggered by the fact the established mechanisms and theories of solidification (nucleation and growth) could be applied to increase the properties just like conventional metallurgical alloys and that this phenomenon was observed in other very important class of materials, that is, Ti-based BMGMC, with much superior casting and mechanical properties. Although research triggered by synchrotron light later on revealed more complex mechanisms of formation (liquid state phase separation prior to nucleation and growth), still, this method of production of BMGMC dominates by far the amount of research activity focused on these class of materials. Investigations such as by high-energy radiation and simulations (Ott et al. 2005; Stoica et al. 2010) just helped to further polish the mechanisms, processes, and/or improve properties of composites.

Credit again goes to Caltech and Ms. Haein Choi-Yim for introducing this concept for the first time in 1997 during her PhD studies (Choi-Yim and Johnson 1997; Choi-Yim 1998), but the real effort came from more permanent members of the group such as Hays, Kim, and Johnson (2000) when they for the first time reported the use of fine dispersed ductile phase crystalline (bcc Ti–Zr–Nb β phase) dendrites in Ti-based BMG matrix (Figure 1.9a) to increase plasticity and study and control their microstructural properties. This was indeed a major discovery as 3D network of equiaxed dendrites produced as a result of controlled solidification and microalloying effectively not only served as a barrier to shear band propagation but also become the reason for their multiplication. This dramatically increased the plasticity, impact resistance, and toughness of BMGMCs.

They improved on their discovery in 2001 (Hays et al. 2001) and it was shown that shear band spacing is coherent with periodicity of dendrites. Mechanical tests were conducted under unconstrained conditions. The generation of shear bands, their propagation and interaction with dendrites is attributed to "free volume" concept rather than dislocation theories, which are major mechanisms of failure in crystalline alloys. At the same time, another group at Caltech studied the effect of shock wave on Zr-based BMG and their newly formed composites (Zhuang et al. 2002). They subjected conventional Vitreloy 1 and new β-Vitreloy 1 to planar impact loading. They observed a unique low-amplitude elastic precursor and bulk wave prior to rate-dependent large deformation shock wave. Spalling was also observed in both Vit1 and β-Vit1 and was attributed to shear localization and debonding of β-phase boundary from matrix at 2.35 and 2.11 GPa (strain rate 2×10^6 s^{-1}), respectively. Since then, the interest in development and improvement of these composites has increased many folds in various groups around the globe. As said earlier, special focus is given to improvement of metallurgical properties related to

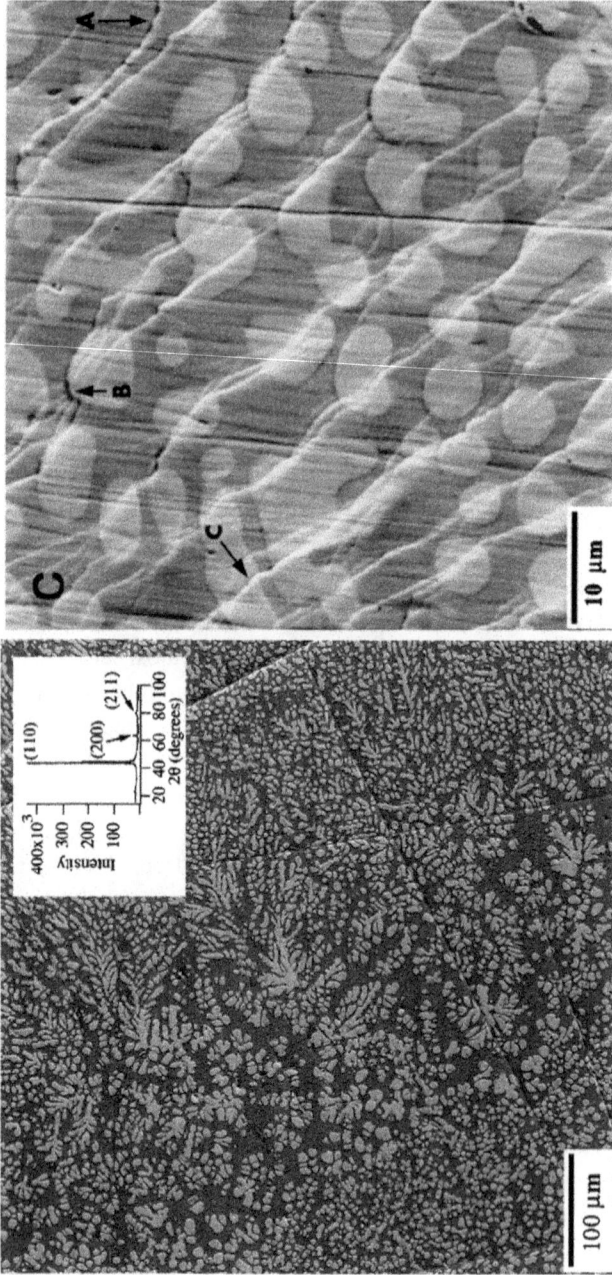

Figure 1.9. (a) SEM backscattered electron image of *in situ* composite microstructure (× 200); (b) shear band pattern array from failed surface showing their crossing dendrites (Hays et al.2000).

thermodynamics and foundry engineering. A notable contribution in this regard is made by a group led by Prof. Jürgen Eckert at IFW Dresden. He and his colleagues have studied various properties of *in situ* BMG composites from various perspectives related to processing, casting (Eckert et al. 2002; Liu et al. 2012; Löser et al. 2004), corrosion, and mechanical. For example, in a study in 2004 (Löser et al. 2004) one of their graduate students (Mitarbeiter), Jayanta Das (now at IIT Kharagpur) studied the effect of casting conditions on mechanical properties of Zr–Cu–Al-Ni–Nb alloys with and without the addition of Nb. It was reported previously that addition of Nb strongly affects the casting properties (fluidity) of this otherwise sluggish and viscous alloy. The research showed that addition of Nb caused a stabilized bimodal size distribution which consisted of bcc β-Zr nanocrystalline dendrites dispersed in amorphous matrix. It was observed that a strong relation exists between the dendrite and matrix crystallite size and morphology, which is a function of casting conditions. The cooling rates ($\varepsilon \sim 2.6 \times 10^3$ K/s to 4.0×10^1 K/s) were estimated from secondary dendrite arm spacing.

Effect of processing parameters on the mechanical properties is less significant for the lower percentages of Zr and Nb, which was increasingly significant for $Zr_{73.5}Nb_9Cu_7Ni_1Al_{9.5}$. Similarly, optimization of properties (reaching maximum of 1,754 MPa fracture stress and 17.5% strain to failure) was only possible toward a rod of Ø 10 mm. Another very important fundamental discovery was the observation of dislocations in these alloys. Although it was observed in dendrites (while shear banding was dominant in bulk glassy matrix), it still was a major breakthrough toward increasing plasticity of these BMGMCs. Figure 1.10 shows optical and electron micrographs of Zr-based *in situ* dendrite/glass matrix composites cast in the form of rods of Ø 5 mm and Ø 10 mm (cold crucible only) by various manufacturing routes. Another important development which took place in their group was the introduction of transformation-induced plasticity (TRIP) BMGMCs. They successfully produced BMGMC in martensitic alloys (Das et al. 2009) and also caused transformation of B2 ductile phase (observable in Zr–Cu–Co systems) to B19′ martensitic phase (with or without twins) (Pauly et al. 2009a, 2010; Javid et al. 2011; Wu et al. 2015a) in this class of material. They are also pioneers of (not quite matured) production of BMGMC by selective laser melting (a form of additive manufacturing; Pauly et al. 2013; Jung et al. 2015b; Prashanth et al. 2015; described in detail in Section 2.3.1.2). Since then various studies have been reported by their group not only in Zr but also Al (Park et al. 2009), Fe, and Ti (He et al. 2003; He et al. 2003; Das et al. 2005a; Gargarella et al. 2013) based BMGMCs in numerous reports (He, Eckert, and Löser 2003; Wu et al. 2006, 2007, 2015a; Eckert et al. 2007; Das et al.

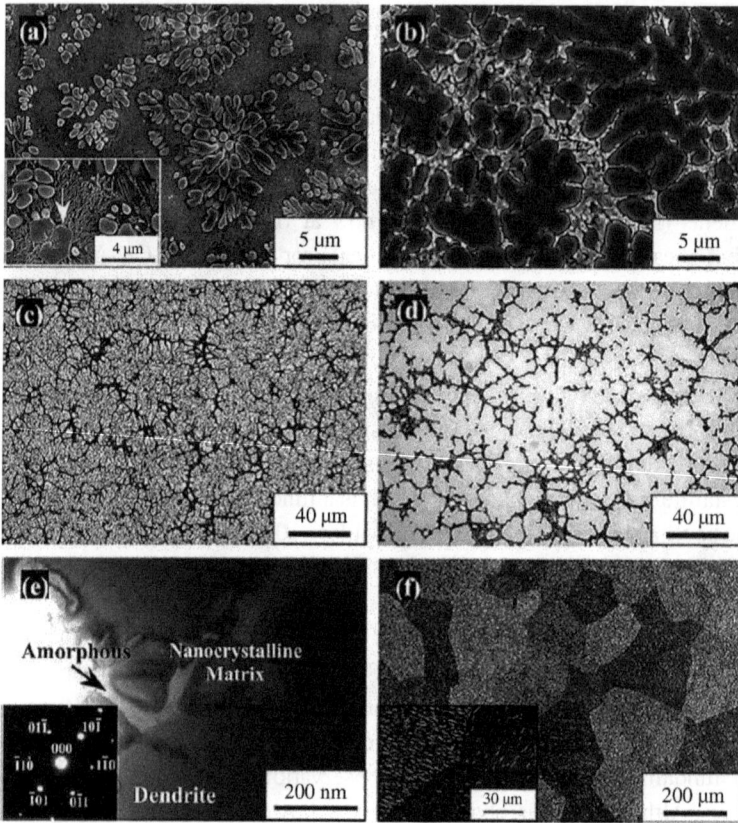

Figure 1.10. Typical microstructures of Zr-based *in situ* dendrite/glass matrix reinforced composites. (a) Injection cast rod (Ø 5 mm) (inset: magnified view of IMC), (b) suction cast rod (Ø 5 mm), (c) centrifugally cast rod (Ø 5 mm), (d) cold crucible cast rod (Ø 10 mm), (e) transmission electron microscopy bright field image of centrifugally cast rod (Ø 5 mm) showing dendrites (inset: SEAD of dendrite) and (f) SEM image of cold crucible cast rod showing various grains (inset: magnified view of precipitates within grains).

2009; Pauly et al. 2009b; Stoica et al. 2010; Song et al. 2011, 2012, 2013; Tan et al. 2011; Qu et al. 2012; Sun et al. 2012, 2013; Wang et al. 2014a; Prashanth et al. 2015) showing improvements in metallurgical, chemical, and mechanical properties of these composites. They are heavily supported by clusters such as DFG, DAAD, EU Marie Curie network, FP7, and Humboldt Foundation with strong collaborative programs with Caltech, JHU, UTK, UST, Tohoku, and IITs.

Another group whose contribution toward development and processing of this class of materials is worth mentioning is led by

Prof. Todd C. Hufnagel at John Hopkins University. They also contributed by fundamental physics (Antonaglia et al. 2014a), microstructural (Fan et al. 2002; Hufnagel et al. 2002a), and mechanical property (Hufnagel 2006, 2013; Hufnagel et al. 2016) improvement. Their contributions led to development of increased plasticity and improved fracture toughness by conventional manufacturing processes. One of the notable discoveries made by their group details the deformation of BMGs by slip avalanches (Antonaglia et al. 2014a). In 2014, Prof. Hufnagel along with his collaborators for the first time reported that slowly compressed BMG deforms via slip avalanche of elastically coupled shear transformation zones which are collection of 10–100 atoms. It was previously believed that deformation in BMG is largely due to rapid movement of shear bands (*which is intermittent instantaneous transience [rearrangement] of atoms*). However, the detailed mechanism about their behavior at low and high temperature and slow and high deformation rates was not known (Schuh et al. 2007). In their study, they reported that slip avalanches are observed at very slow compressive deformation rates only in accordance with serrated flow behavior. A schematic of setup used and patterns observed are shown in Figure 1.11. At very high temperatures, there is no problem in explaining the behavior of BMGs which is primarily homogeneous. However, this does not remain same at low temperatures and changes to inhomogeneous type which is characterized by onset of intermittent slips on top of narrow shear bands. If this inhomogeneous deformation is coupled with low strain rates, it gives rise to sudden stress

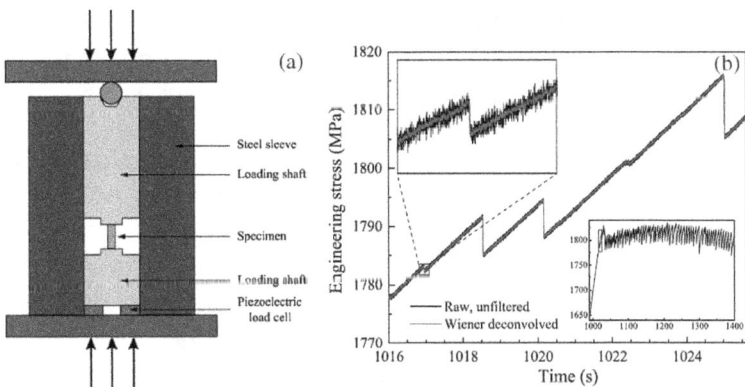

Figure 1.11. (Left): Schematic diagram of experimental setup. (Right): Lower right inset: Stress vs time. Main figure: Magnification of data in small window in lower right inset. Upper left inset: Magnification of stress drop (serration) during slip avalanche. (Black lines: Raw unfiltered stress vs time. Red Lines: time series after Wiener deconvolution) (Antonaglia et al. 2014a).

drops called serrated flow which was found to be consistent with earlier reports (Georgarakis et al. 2008; Qiao et al. 2010; Sun et al. 2012) (which is a new discovery). This discovery led to better understanding of behavior of BMGs and consequently provides a basis for the manufacture of strong and tough materials.

Elaborative details of their work could be found in reports published from their group (Hufnagel et al. 2002a, 2002b; Ott et al. 2005, 2006; Hufnagel et al. 2006; Schuh et al. 2007; Cunliffe et al. 2012; Hufnagel 2013; Antonaglia et al. 2014a; Hufnagel et al. 2016). There are further more notable contributions from the United States (e.g., Prof. Peter K. Liaw at UTK [primarily about fatigue, shearing, and nano-indentation] [Fan et al. 2006a, 2006b; Jiang et al. 2006; Du et al. 2007; Miller and Liaw 2007; Jiang et al. 2008; Qiao et al. 2009a, 2013a, 2013b, 2013c; Wang et al. 2009; Chu et al. 2010, 2012; Wang et al. 2010c; An et al. 2012; Antonaglia et al. 2014b], Prof. Jan Schroers at Yale), Taiwan (Huang et al. 2009; Qiao et al. 2016), and Korea (Jeon et al. 2012; Kim et al. 2013; Rim et al. 2013; Jung et al. 2015a; Yi et al. 2016) which provided footstone for the further development and property enhancement of this class of materials.

1.8.2 SOLID–SOLID (S–S) TRANSFORMATION (DEVITRIFICATION)

A remarkable property exhibited by BMGMCs is that after being casted in vitrified shape, they tend to relax when subjected to the effect of heat. This is very much similar to phenomena observed in monolithic BMGs which accounts for their ductility. This gives rise to another mechanism by which they could be formed with enhanced strength and toughness known as devitrification or solid state phase transformation. This is primarily attributed to either of two mechanisms which may be responsible for the onset of crystallization upon heating. These are known as "solid state phase separation" (Fan et al. 2006b) and "quenched in nuclei" (Antonione et al. 1998; Krämer et al. 2015). Solid state phase separation also known as "spinodal decomposition" in conventional alloys (Porter and Easterling 1992) is separation of phases which occurs before the growth of second phase in BMGMCs. This is not nucleation and growth phenomena but solid state detachment based on miscibility of two phases with each other. Although observed in selected compositions only (Van De Moortèle et al. 2004; Oh et al. 2005; Park and Kim 2006; Park et al. 2006; Park, Kyeong, and Kim 2007; Kim et al. 2013), this accounts for a unique mechanism for the formation of ductile dendrite-reinforced BMGMCs

(Chang et al. 2010; Bracchi et al. 2006). Second mechanism which is known as "quenched in nuclei" is an intermittent metastable nuclei of ductile phase dendrites which have just started to grow but their growth is suppressed by rapid cooling to which they are subjected to quickly thereafter. This retards any further growth. However, as they are metastable and still in their high-energy state with the drive to transform to fully grown stable phases, they fulfill their desire when subjected to heat and serve as sites for the growth of full-scale second-phase crystalline dendrite. This is a very important mechanism of fabricating these kinds of composites. These are rather well-established mechanisms and account well for the phenomena observed.

1.9 PRODUCTION METHOD—PROCESS PERSPECTIVE

From a process point of view, production methods could be broadly classified into five main methods used for the fabrication of BMGMCs till date. These are melt spun, twin roll casting (TRC), Cu mold suction casting, semisolid processing, and additive manufacturing. This classification by any means is not exhaustive but narrates only those methods which are easy, cheap, reproducible, versatile, and have been used repeatedly over years for the manufacture of ductile glassy composites. Their brief details are as follows.

1.9.1 TWIN ROLL CASTING

This, by far, is the first method ever used for the manufacture of glassy alloys. This is a variant of melt spun method which is responsible for the first observations of glassy structure in metals by Paul Duwez and coworkers (Klement et al. 1960) as they casted thin ribbons of Au–Si alloy which were first to retain glassy structure at room temperature. The process, in essence, consists of two high-speed self-driven water-cooled Cu rolls rotating in opposite direction, which are finely ground and polished to avoid any contamination and variation in thermal gradient across surface caused by impurities. Furthermore, the wheels are constantly replenished with lubricating medium (usually organic surfactants) (if necessary) to ensure their smooth functioning and near-friction free rotation. This and other processes have been excellently documented by Prof. Inoue in a recent article (Inoue et al. 2015).

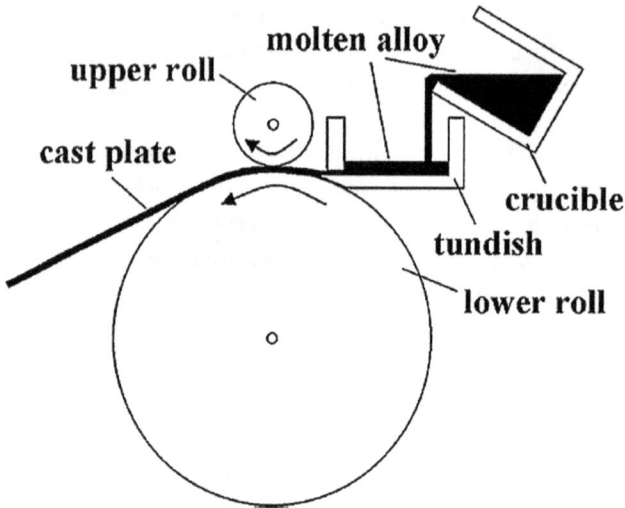

Figure 1.12. Schematic TRC setup (Inoue et al. 2015).

TRC underwent a considerable hierarchical development from earlier variant of wire caster to rotating disk caster to finally its TRC form (Figure 1.12).

Modeling and simulation of the process and the solidification microstructure evolved in it has also been carried out (Wang et al. 2010a) to further optimize the processing conditions and set process parameters. The process is excellent for making thin continuous lengths of ribbons of almost all compositions since process is not affected by any change in alloy composition. Chilling effect is caused by the very nature of water-cooled copper rolls, their surface area, and rotation. Water temperature is usually kept at 8°C and speed of roll is variable (could be adjusted for maximum efficiency). Copper is selected as the material for rolls as it has excellent thermal conductivity; thus, it can easily conduct heat away from it. Alloy is poured from ladle or crucible to a tundish that transfers the melt to rotating copper rolls. It is by far the first and still the only mass production method for making BMG and their composites.

1.9.2 CU MOLD (SUCTION) CASTING

As the conductivity of copper is maximum and it is a very good material for heat transfer, it paved the way toward an idea of using it as mold to make ingots rather than ribbons. The chilling effect is caused by water cooling coupled with suction caused by difference of pressure at the bottom of mold (creation and destruction of vacuum) to that at top. This results in bulk shapes of BMGs. This was, and still is, the most frequently used method for the production of 3D BGAs. The process essentially is the melting of BMG in a small arc melting button furnace. This is followed by the introduction of melt over a nozzle type opening which is controlled by a suction duct/ chute attached to a vacuum pump at the bottom. The pipe connected to vacuum pump is attached to inert gas (Ar) cylinder at the same time with the help of three-way rotating valve. Once the alloy is in its molten state over suction chute (nozzle), three-way valve is quickly shifted toward vacuum pump which sucks all the argon present in chamber through that chute, thus taking away the alloy with it as well. The alloy along its way toward vacuum pump encounters water-cooled chilled Cu mold (plate or cylinder), thus it quickly gets solidified just like any conventional casting process. Once casting is completed, the valve is turned toward air and setup is allowed to cool a little before opening (mold break out/fettling). This is a very efficient, quick method for the production of BMG and their composites and again nearly all types of compositions could be casted as process is composition independent. There have been different approaches in the way different groups have been doing the castings using this technique (Inoue and Zhang 1995, 1996; Wall et al. 2006; Figueroa et al. 2007; Qiao et al. 2009b; Lou et al. 2011), but the fundamental principle remains the same. A recent approach adopted at Huazhong UST, Wuhan (Zhu et al. 2014) is described in Figure 1.13.

1.9.3 SEMISOLID PROCESSING

Semisolid processing, most popularly tried in its variant forms such as "semisolid forging," is another innovative technique for processing BMGMCs. Essentially, in this process, alloy is melted by taking the composition to its melting temperature and then upon cooling first it is held in a region in between liquidus and solidus lines in which nucleation and growth of second phase particles (dendrites) is fully completed. Following this, the alloy is rapidly quenched to room temperature. This "hold" in between liquidus and solidus ensures the formation and homogeneous distribution of ductile phase equiaxed dendrites, which are necessary for the development of ductility and toughness. The process is unique in a sense that alloy stays in semisolid state when it is going through primary liquid to solid transformation (solidification). The hardness of material in this region is low and it is possible to impart external forces on it causing its deformation. This is called "semisolid" processing. From a practical viewpoint, it is the only method of fabrication of semicomplex shapes from BMGMCs. Researchers at Caltech in parallel with UC San Diego in 2009 (Hofmann et al. 2009; Khalifa et al. 2009) used this technique to devise a novel method

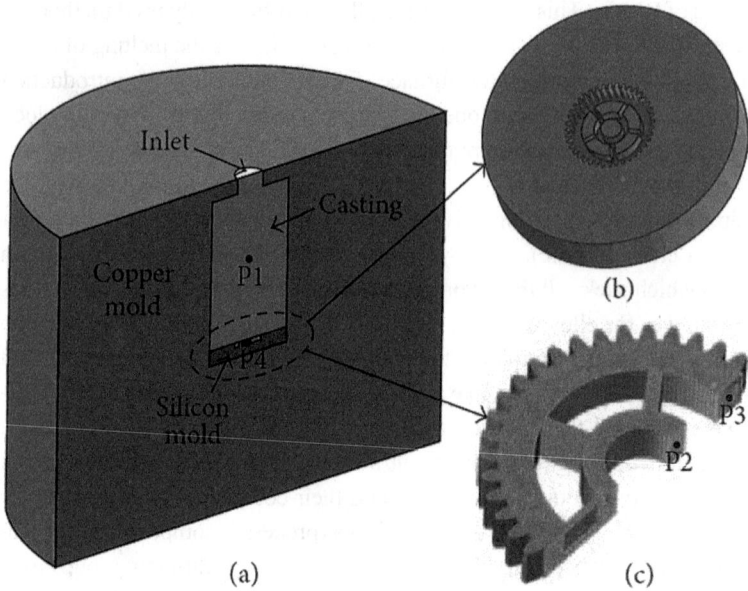

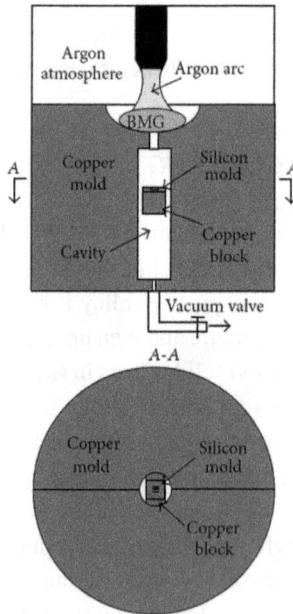

Figure 1.13. (Top) (a) 3D view of casting assembly (b) and (c) expanded view of silicon mold for casting of gear; (Bottom) (Upper) 2D side and (Lower) top view of micromold for suction casting (Zhu et al. 2014).

employing "double boat" and fabricated hemispherical shapes out of different compositions. This was by the application of force using a stub (named as forging). The process is described in detail in Figure 1.14(a–d).

The same technique and its extended version were used for microgravity experiments on board NASA Space Shuttle Program during the 1990s in three notable missions: STS-65, STS-83, and STS-94 (Hofmann and Roberts 2015). There were other efforts following this as well which were based on same principle but the technique was extended to make complex shape by so-called "super plastic forming" (Wang et al. 2009a) or "TPF" (blowing [Schroers 2010; Sarac et al. 2011]) (Schroers 2005; Duan et al. 2007; Jeong et al. 2009; Wang et al. 2009a; Li et al. 2016; Qiao et al. 2016).

1.9.4 ADDITIVE MANUFACTURING

This is the most recent and advanced process of making BMGMCs (also the subject of the present study). Various efforts have been made to produce Al (Li et al. 2014b, 2014c, 2015; Olakanmi et al. 2015; Prashanth et al. 2015), Fe (Zheng et al. 2009; Kruth et al. 2004; Jung et al. 2015b), Ti (Thijs et al. 2010; Zhang et al. 2011b; Dutta and Froes 2016; Vastola et al.

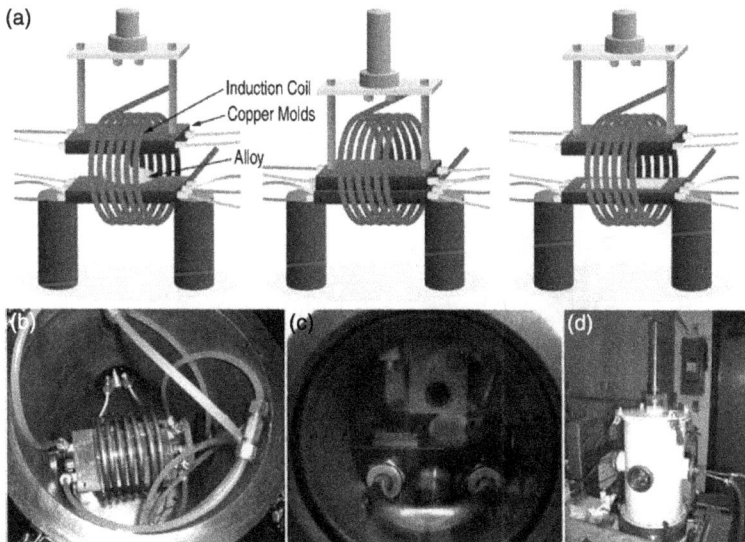

Figure 1.14. (a) Schematic of double boat method. (b) Photographs of mold together inside the casting chamber; (c) casting chamber closed; (d) external full view of setup showing inlets and cooling water (Hofmann et al. 2009).

2016), and Zr (Santos et al. 2006; Sun and Flores 2008, 2010, 2013; Chen et al. 2010; Yang et al. 2012; Welk et al. 2014; Zhang et al. 2015)-based BMGMCs by different variants (selective laser sintering [SLS], selective laser melting [SLM] [Pauly et al. 2013]/laser engineered net shaping [Borkar et al. 2016], powder feed deposition [PFM]/direct laser deposition [DLD] [Shamsaei et al. 2015; Thompson et al. 2015], wire feed method [Ding et al. 2015], electron beam method [Romano et al. 2015]) of additive manufacturing (AM; Santos et al. 2006; Frazier 2014; Sames et al. 2016). The process essentially consists of supplying material (BMGMC) in the form of powder coaxially or at an angle to vertical axis (PFM/DLD), or wire (WFM) just like welding (Metal Inert Gas/Gas Metal Arc Welding), or making shape out of powder bed by selectively sintering (SLS)/ melting (SLM) parts by the impingement of laser on the bed. The motion of laser is controlled by the robotic arm which is controlled by a microprocessor and this motion is a function of coordinates given to controller (computer) in the form of CAD model. There are variants in which CAD model could be generated as a part of whole setup, is sliced by a slicing program (which divides the geometry of part into layers [layer-by-layer (LBL) fashion]), and is given to final laser controlling robot which dictates its motion as LBL pattern forms (Figure 1.15). *Advantages* of the process are (a) it is a one-step direct fabrication of composite parts in near

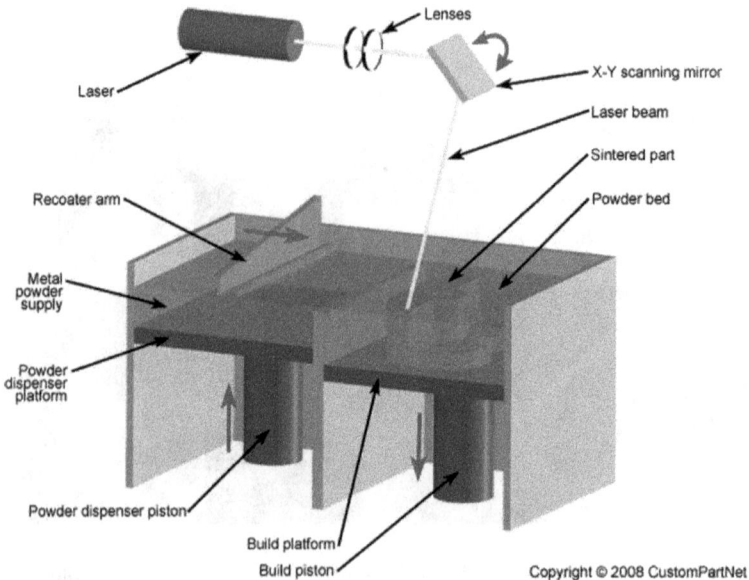

Figure 1.15. Schematic of SLM systems (Sames et al. 2016).

net shape, (b) there is no need for any posttreatment (finishing, deburring, and heat treatment) facilitating the attainment of properties directly without incurring additional costs, (c) surface finish of parts produced is excellent, (d) excellent dimensional accuracy could be achieved, and finally (f) excellent microstructural control could be exercised in one step. *Limitations* are it is still a new process and optimization of properties as a function of laser parameters is a challenge, (b) reproducibility is difficult from laboratory to laboratory, (c) almost nothing is known about metallurgy of melt pool due to its incipient and transient nature, thus mechanisms of nucleation and growth coupled with heat transfer are not known (aim of present study), (d) powders for AM require special production procedures and are expensive, (e) process is largely dedicated to repair (including cladding) rather than full-scale part production due to aforementioned reasons.

Notable discoveries have been made by groups at Ohio State University (OSU) led by Prof. Katherine Flores (now at Washington University St. Louis) (Zr-based BMGMC) (Sun and Flores 2008, 2010, 2013), Dr. Brian Welk (OSU) along with Dr. Mark Gibson (CSIRO, Australia) (compositionally gradient alloys [BMGMC–High-Entropy Alloys]) (Borkar et al. 2016; Welk et al. 2016), Prof. Tim Sercombe at the University of Western Australia, Perth (Al-based BMGMCs) (Li et al. 2014b, 2014c, 2015), Harooni and colleagues at the University of Waterloo, Canada (Zr-based alloys) (Harooni et al. 2016), Prof. Scott M. Thompson at Mississippi State University (MSU) (Shamsaei et al. 2015; Thompson et al. 2015), Prof. Jurgen Eckert at IFW Dresden (Zr-, Al-, and Fe-based BMGMCs) (Pauly et al. 2013; Jung et al. 2015b; Prashanth et al. 2015), Wang and colleagues in Taiwan (Zr-based BMGMCs) (Wang et al. 2010), Yue et al. (Zr-based BMGMCs) (Hong Kong) (Yue et al. 2007; Yue and Su 2008), and Prof. Huang at Northwest Polytechnic University, China (Zr-based BMGMCs) (Yang et al. 2012; Zhang et al. 2015).

Advantages associated with the use of AM techniques in the processing of BMGMC are in addition to above-mentioned advantages of process as a whole. For example, in AM of BMGMC, extremely high cooling rate inherently present in the process gets exploited in a sense that it helps in bypassing the nose of TTT diagrams of composites, thus enabling formation of glassy structure without "phase separation (Liquid Liquid Transition) (Zu 2015)" or "Nucleation and Growth" of crystalline phases (Pauly et al. 2013). The same high cooling rate and glassy phase formation is responsible for imparting very high strength in these systems. Similarly, LBL formation helps in development of glassy structure in top layer and as the laser travels the path dictated by geometry of part to be produced, the preceding layer get heated to a temperature somewhere inside the nose of TTT diagram. This causes devitrification and if controlled properly in a narrow window of time and composition, gives rise to tough and strong structure in one-step fabrication. The same process is limitation as well, as when composite (i.e., layer beneath fusion layer) spends too much time inside nose of TTT diagram, it may cause complete (100%) crystallization forming crystalline alloy which is neither desirable nor is the source of enhanced properties of laser-processed BMGMCs. Latest trend is to use compositionally gradient alloy compositions (Hofmann et al. 2014a, 2014b) along with advantages of AM to fabricate a series of alloys which have a spectrum of properties as a function of composition and laser parameters (power, scan depth, scan rate, and spot size). The final composition which is best and suitable for a particular application could be chosen thereafter depending on type of application. This gives rise to not only production of BMGMCs but also incorporation of HEAs (Joseph et al. 2015; Ocelík et al. 2016) in this research. Very recent publications by Dr. Mark Gibson (Borkar et al. 2016; Welk et al. 2016) are notable in this regard. Laser deposition has also been used to identify GFA (an age-old disputed property of glassy systems) (Tsai and Flores 2015).

1.10 EQUILIBRIUM PHASE DIAGRAMS

BMGs traditionally were explained by the help of equilibrium diagrams, mainly binary in nature (Miracle et al. 2010). As in early days (before the establishment of three laws in last decade of last century) only binary compositions were rapidly quenched to produce glassy structure mainly for laboratory purposes and no efforts were made to understand the exact, complex, and varied behavior of these alloy systems out of laboratory conditions on industrial scale. It was only after 2007 (establishment of three rules) (Hofmann and Johnson 2010; Inoue and Takeuchi 2011) that ternary diagrams were more often used to predict and understand the behavior, in particular GFA of this class of materials. Most of those diagrams were based on eutectic compositions despite the fact the GFA is most often off-eutectic and is merely an empirical parameter (as discussed in Section 2.4.1). These diagrams are helpful to a certain extent only in understanding the behavior of BMGMCs. Since behavior of BMGMCs in actual conditions is not at all equilibrium cooling due to rapid quench, more effective way to understand and explain this is by means of nonequilibrium or TTT diagrams (explained in the next section).

1.11 NONEQUILIBRIUM PHASE DIAGRAMS (E.G., TTT DIAGRAMS)

Since most of the BMGMCs do not form under slow cooling and in particular very high cooling rates are met in AM, their behavior and development of phases could best be explained by nonequilibrium, for example, TTT diagrams (Figure 1.16). These diagrams are very similar to TTT diagrams used extensively in steel metallurgy. The only major difference is their nose is shifted to extreme right because of their multicomponent nature (Schroers 2010). Like other alloys, their behavior is dictated by alloying elements present in them. Most of the alloying elements shift the diagram to right, making it easy for the glass to form. These are also helpful in calculating GFA (Zhu et al. 2007a). They are also very helpful in explaining and designing of procedures for Thermoplastic forming (TPF) of these alloys (Pauly et al. 2013).

1.12 INVARIANT TEMPERATURES (T_L, T_m, T_g, T_x)

Referring back to Figure 1.16, distinct regions on TTT diagram can be identified. This includes invariant temperatures marked by T_L (liquidus

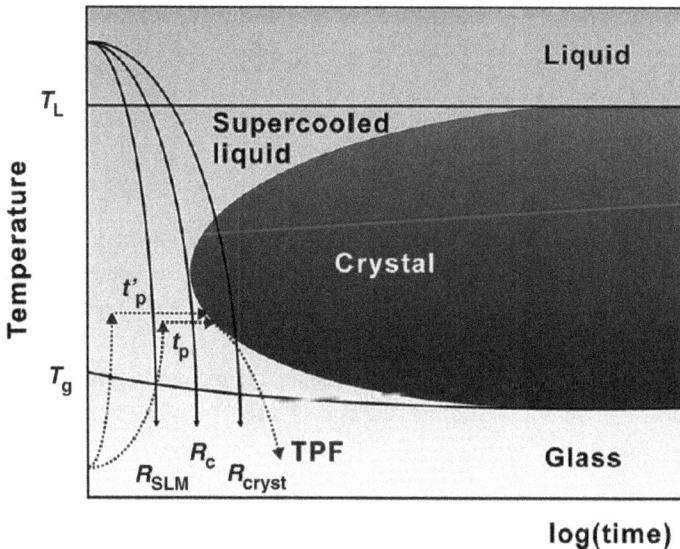

Figure 1.16. Schematic TTT diagram for BMGMCs (indicating regions [cooling rates]) of monolithic glass formation as well as TPF (Pauly et al. 2013).

temperature [upper line]), T_m (melting temperature [not shown and is usually equal to liquidus temperature for most alloys]), T_s (solidus temperature [not shown and is usually visible on equilibrium phase diagram, temperature at which a phase transformation from liquid to solid is completed]), T_g [glass transition temperature [lower line]), and T_x (crystallization temperature [onset of crystallization, not shown]). There is additional information available and can be used from TTT diagrams. This includes cooling rates data. For example, above figure shows R_c (which is critical cooling rate exactly necessary to form fully glassy [monolithic] structure), R_{cryst} (cooling rate for crystallization). This is the rate at which, if an alloy is cooled, it results in crossing the nose of TTT diagram resulting in the formation of partial or full crystals in glassy matrix (not desirable), R_{SLM}. (This is critical cooling rate usually desirable, and is achieved, in SLM. It clearly shows that no crystallization will happen and fully glassy structure will form.)

1.13 COMMON MICROSTRUCTURES

Although a function of alloy composition to a large extent, this section details the microstructures commonly observed in Zr based as cast hypoeutectic (Zr > 65%) and eutectic (Zr < 50%) systems used in this study. The alloys investigated are $Zr_{47}Cu_{45.5}Al_5Co_2$ (eutectic) and $Zr_{65}Cu_{15}Al_{10}Ni_{10}$ (hypoeutectic). Their microstructures are explained below.

1.13.1 $Zr_{65}Cu_{15}Al_{10}Ni_{10}$ SYSTEM

This system primarily consists of

a. Zr_2Cu type tetragonal phase formed at very high cooling rates only and
b. Zr_2Cu + eutectic (Zr_2Cu + ZrCu) type phase which is formed at intermediate (6 mm/sec) to slow (1 mm/sec) cooling rates

An inverse relation exists between eutectic and cooling rate. Amount of eutectic increases as cooling rate is decreased.

$$eutectic \propto \frac{1}{cooling\ rate} \qquad (5)$$

Other phases which are present in these alloys are τ_3 and τ_5. However, these are not observed as there is Ni in the system replacing some of Cu. Second prominent effect which is observed in these systems is the effect of

Zr content. Table 1.1 shows Zr content and its effect on phase development at constant withdrawal velocity of 6 mm/sec.

Third important observation in this class of alloys is evolution of percentage of crystalline phase, its morphology and percentage of glassy matrix with cooling rate (expressed in terms of withdrawal velocity). This is elaborately explained in Table 1.2.

This also confirms the relation observed in equation (4). In addition, in this class of alloys invariant temperatures have been observed to have

Table 1.1. Qualitative analysis of different phases present in Zr–Cu–Al–Ni Alloy system (Cheng and Chen 2013)

Sr. No.	Zr content	Crystalline precipitates	Glassy substrate
1	Zr_{57}	Zr_2Cu type (similar to Zr_{60} [tetragonal] but different in morphology)	(in percentage)
2	Zr_{55}	Zr_2Cu type (similar to Zr_{60} [tetragonal] but different in morphology)	(in percentage)
3	$Zr_{52.8}$	Nil	(100%) (monolithic BMG)
4	$Zr_{50.1}$	ZrCu type (monoclinic)	(in percentage)

Table 1.2. Qualitative analysis of effect of cooling rate on evolution of different phases (Cheng and Chen 2013)

Sr. No.	Cooling rate (mm/sec)	Crystallite (percentage)	Morphology	Glass (percentage)
1	6	Nil	Nil	100%
2	4	Zr_2Cu I ZrCu eutectic (<100%)	Spherical	<100%
3	3	Zr_2Cu + ZrCu eutectic (<100%)	Spherical	<100%
4	1	100% Zr_2Cu + ZrCu eutectic	Spherical	Nil

following behavior. Glass transition temperature (T_g) is observed to have inverse relation with Zr content (Figure 1.17(a)).

Note that Zr_{55} is at $T_L = 1,157$ K which is eutectic temperature. However, $Zr_{52.8}$ is best glass forming composition which is off-eutectic. This is a contradiction. However, it is an empirical relation and experimental results indicate that ΔT_x, T_{rg}, and/or γ do not best express GFA in these systems. This is typical case of presence of best GFA at off-eutectic temperature as is witnessed by earlier observations (Ma et al. 2003). Similar behavior is observed previously for some Cu- and La-based BMGMCs. However, more research (e.g., variation of percentage of ductile phase and its number density and its relationship with GFA) is needed to verify this hypothesis in hypoeutectic Zr-based systems. Another important fact observed in these systems is effect of variation of GFA with Nb content. Nb is observed to have very prominent effect on fluidity and mechanical properties as controlled by tuning of microstructure in these alloys (Fan et al. 2001; Kühn et al. 2004; Sun et al. 2005b). For example, in a study conducted by Sun et al. (2005) it was shown that addition of Nb up to maximum of 15 at% causes precipitation of β-Ti-like dendrite phases in glassy matrix. These dendrites are few in number at 5 at% and tend to increase with increasing Nb content with the formation of other Quasi Crystalline (QC) particles. Their behavior is qualitatively shown in Table 1.3.

This study confirms their observations in other similar efforts aimed at tuning other properties by controlling dendrite parameters (type, size, and shape) and microstructure (Sun et al. 2005a, 2005c). It is also observed in another study by Prof. Inoue and colleagues that crystallization process of Zr–Ni–Cu–Al MG is greatly influenced by adding Nb as an alloying element (Fan et al. 2001). Based on the results of the DSC experiments for MGs $Zr_{69-x}Nb_xNi_{10}Cu_{12}Al_9$ ($x = 0$–15 at%), the crystallization process takes place through two individual stages. For ($x = 0$), metastable hexagonal ω-Zr and a small fraction of tetragonal Zr_2Cu are precipitated upon completion of the first exothermic reaction. The precipitation of a nano-QC phase is detected when Nb content is raised to 5–10 at%. Similar trends were observed in studies conducted by Prof. Eckert's group at IFW, Dresden (Kühn et al. 2002, 2004). The ongoing research on this class of materials shows and tally with the observations made earlier proving grounds for the validity of hypothesis that nucleant serve as sites for copious nucleation of ductile phase dendrites (Song et al. 2016).

1.13.2 $Zr_{47}Cu_{45.5}Al_5Co_2$ SYSTEM

This is the system in which not only famous ductile phase B2 bearing regular bcc structure is observed but its transformation product B19′

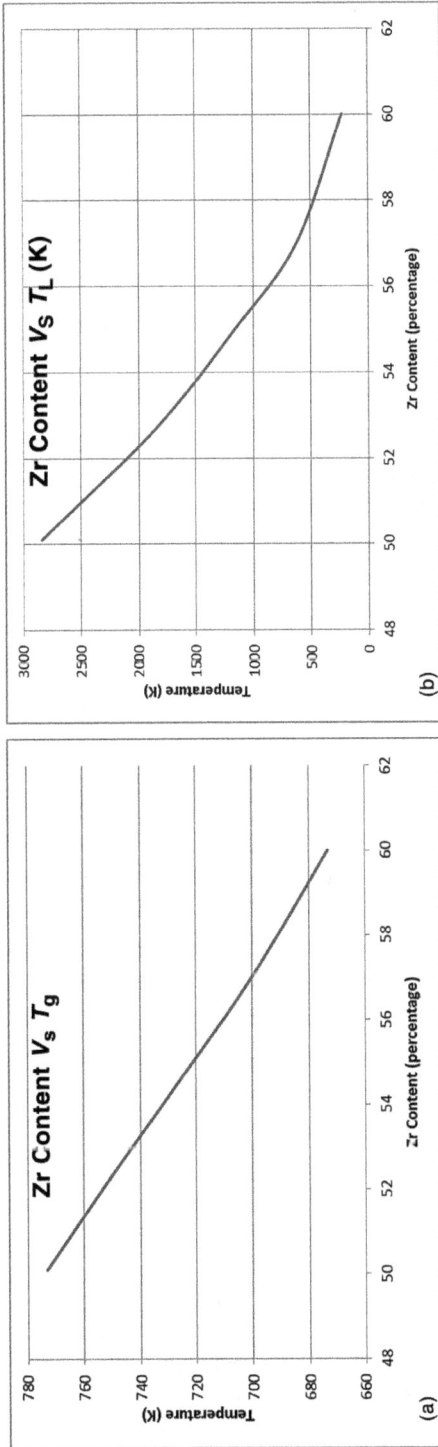

Figure 1.17. (a) Graphs showing relation between glass transition temperature (T_g) and Zr content. (b) Graphs showing relation between liquidus temperature (T_L) and Zr content (Cheng and Chen 2013).

T_x = crystallization temperature (onset of crystallization) is independent of composition.

T_m = Melting temperature is constant for all alloys at 1,094 K, indicating that all alloys are formed at same constant eutectic reaction temperature.

T_L = Liquidus temperature shows nonlinear (decreasing trend) dependence on composition (Figure 1.17(b)).

Table 1.3. Qualitative analysis of effect of Nb content on evolution of different phases and ultimate fracture strength (K_{1C}) (Cheng and Chen 2013)

Sr. No.	at% Nb	β phase dendrites	QC particles	Ultimate fracture strength (K_{1C}) (MPa)
1	5	Low percentage (<100%)	Nil	1,793
2	10	Intermediate percentage (<100%)	<100%	1,975
3	15	High percentage (<100%) (fully grown 3D morphology)	<50%	1,572
				.

(bearing martensitic structure) is also observed (Gao et al. 2015). In these ZrCu-based alloys, strain hardening rate is enhanced and plastic instability is suppressed due to a martensitic transformation of B2 to B19'. In fact, shape memory effect (which is evolution of unique property [Biffi et al. 2013]) is also observed due to simultaneous reversible deformation of strained B19' (present in certain percentage) along with a certain percentage of regular strain free B2 ZrCu bcc. The presence of these two fractions cause a tuning effect which gives rise to shape memory phenomena (i.e., strain free lattice [regular bcc] can be reversibly changed to strained lattice [martensite] by the application of heat, causing restoration of shape [Nishida et al. 1998; Firstov et al. 2004; Schryvers et al. 2004]). The detailed mechanism for a system studied by Wei-Hong and coworkers (Gao et al. 2015) is given below. Shape memory effect along with GFA is associated with martensitic transformation of B2 to two monocline structures.

a. A base structure (B19') with $P2_{/m}$ symmetry and
b. A superstructure with C_m symmetry

Transformation temperature hysteresis of ZrCu-based shape memory alloy is large while thermal stability is poor. Grain size is observed to have inverse relation with Co content (expressed in percentage). Average grain size of $Zr_{47}Cu_{45.5}Al_5Co_2$ is 6 μm. The microstructures observed in these alloys are Co_2Zr_3 and B2. Transmission electron microscopy shows that both austenite and martensite coexist, which is an indication of the fact that Co ensures the stability of martensite over a large temperature range. In other words, martensite transformation temperature becomes low. This martensite exists in C_m symmetry.

Note: Rietveld refinement shows that

a. At normal conditions: in intermetallic compounds (IMCs) $Zr_{50}Cu_{50}$, two types of martensite exist, namely B19' and C_m. Both have certain volume fraction present in conjunction with each other. B19' has 27% V_f while C_m has 73% V_f.
b. Under compositional contract conditions:
   a. When the content of Al atom substituting for Zr atom is smaller than 9.375% (mole fraction) (Al < Zr 9.375%): Austenite phase could form a martensite base structure during quenching or straining (popularly known as stress-induced martensitic transformation/TRIP). This phenomenon is observed not only in Zr–Cu–Al–Co systems but also in many other systems (Seo and Schryvers 1998a, 1998b; Lee et al. 2004; Shi and Falk 2007; Louzguine-Luzgin et al. 2009; Liu et al. 2010a; Pauly et al. 2010; Wu et al. 2010; Kim et al. 2011; Hao et al. 2013; Song et al. 2013).
   b. When Al > Zr 9.375%: Austenite phase could form a super-structure (C_m)
c. Co-doing: Another important phenomenon is "Co-doping" of Al and Co. This reduces the formation of B19', thus makes it even more difficult to find B19' martensite.
d. "One-step" transformation: Another notable observation is that only "one-step" transformation occurs, that is, B2 transforms directly to C_m. Only one exception is $Zr_{47.5}Cu_{46.5}Al_5Co_1$ in which case B2 first transforms to B19' and then B19' transforms to C_m phase upon cooling. In this case, $M_s = 309$ K while $M_f = 275$ K.

Addition of aluminum causes a decrease in martensitic transformation temperature (M_f) until the Al content reaches a value slightly greater than 6%. However, M_s remained almost constant. *Addition of cobalt (Co)* M_s temperature rapidly decreases with addition of Co content. When the addition of Co increases to 2%, the martensitic transformation temperature (M_s) and transformation hysteresis change invariably. This happens as a result of variation of intrinsic factors, that is,

a. Increase in unit cell volume and
b. Decrease in electron concentration with increasing Co content (because Co has small atomic radius and high electron concentration).

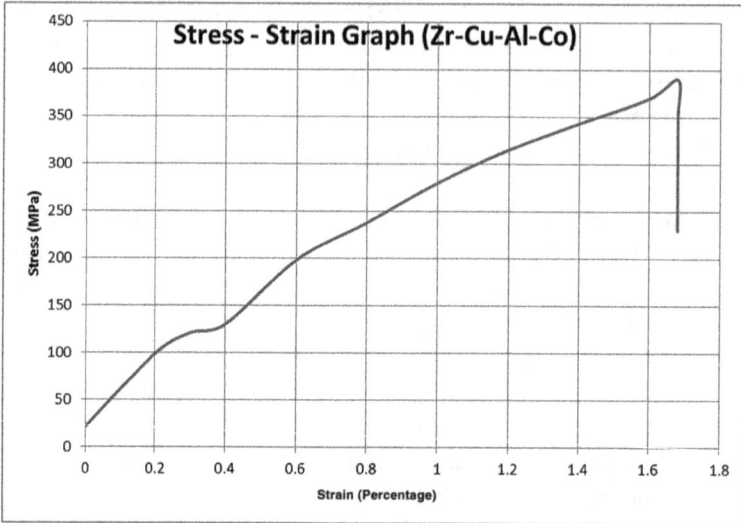

Figure 1.18. Stress strain graph of $Zr_{47.5}Cu_{45.5}Al_5Co_2$ (Gao et al. 2015).

Mechanical properties: Stress strain curve of $Zr_{47.5}Cu_{45.5}Al_5Co_2$ is shown in Figure 1.18.

Compressive strength of alloy increases with increase of Co content. This is attributed to shear-induced martensitic transformation from cubic B2 to a monoclinic martensite phase (C_m), which imparts an appreciable work hardening capability. *Fracture strain* increased from 0.73% to 1.76% as Co content varies from 0.5% to 2%. Fracture surface analysis revealed that at lower concentrations, intergranular fracture dominates. As cobalt content increases to 2%, ductile fracture features started to appear. Surface at this concentration is characterized by a lot of faults and tearing ridges which is indicative that plastic deformation happens first prior to failure. The addition of Al and Co significantly refines grains. The martensite plates become fine. The substructure of alloy is mainly (001) compound twin and martensitic variants are (021) type 1 twin related. In a microstructure of fractured surface observed under scanning microscope, charged surface indicates fracture features (Figure 1.19). (Note: From a crystallographic viewpoint, B2 is cubic in nature, while B19′, in its both morphologies [i.e., $P_{m/2}$ and C_m] is monoclinic).

1.14 MECHANICAL PROPERTIES

Like microstructure, mechanical properties of BMGMC are a strong function of composition. Due to the alloy systems under investigation, a

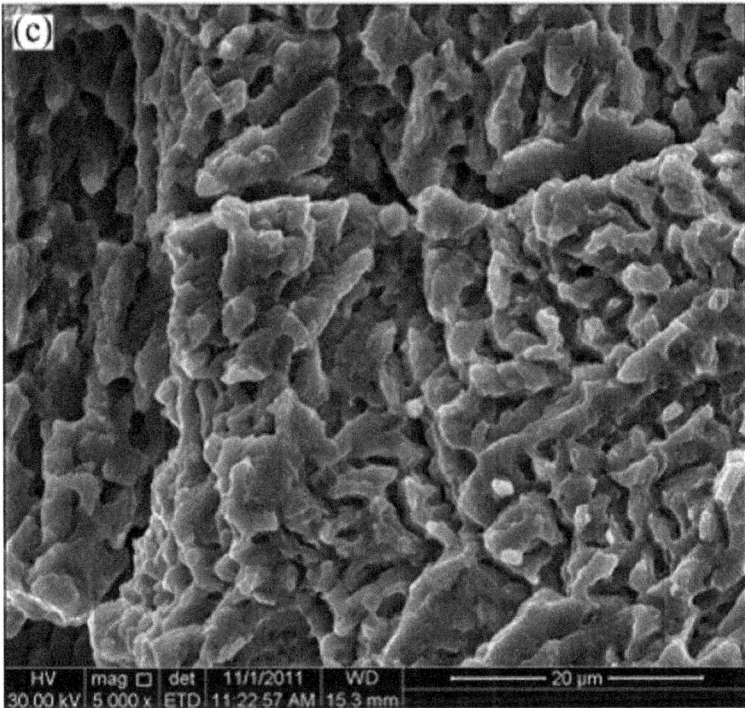

Figure 1.19. SEM image of fracture surface of $Zr_{47.5}Cu_{45.5}Al_5Co_2$ (Gao et al. 2015).

Table 1.4. Mechanical properties of different ZrCu-based eutectic systems (Gao et al. 2015)

Sr. No.	Alloy	Yield stress ($\sigma_{0.2}$ (MPa))	Maximum stress (σ_b (MPa))	Fracture strain (δ/%)
1	$Zr_{48}Cu_{47.5}Al_4Co_{0.5}$	136.25	181.08	0.73
2	$Zr_{47.5}Cu_{46.5}Al_5Co_1$	275.84	311.82	0.75
3	$Zr_{47.5}Cu_{45.5}Al_5Co_2$	367.95	392.59	1.76

contrast, as observed in their varied composition, is described only here. For example, $Zr_{47.5}Cu_{45.5}Al_5Co_2$ in Table 1.4 shows the 0.2% offset yield stress ($\sigma_{0.2}$ [MPa]), maximum stress (UTS) (σ_b [MPa]), and fracture strain (δ/%) of different compositions of aforementioned alloy.

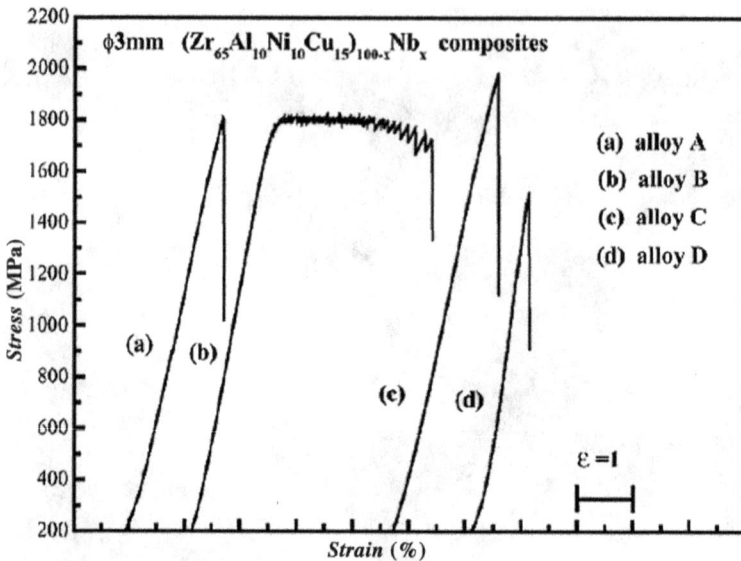

Figure 1.20 Room temperature compressive stress–strain curves of as cast $Zr_{65}Cu_{15}Ni_{10}Al_{10}$ with different percentage of Nb (Alloy A [Nb = 0 at%], Alloy B [Nb = 5 at%], Alloy B [Nb = 10 at%], Alloy C [Nb = 15 at%]) (Sun et al. 2005b).

Similarly, Figure 1.20 shows the compressive stress–strain curves of different compositions of ZrCuAlNi alloy with and without Nb content at room temperature (maximum till 15 at%).

It shows a dramatic behavior of change in yield stress, maximum stress, and fracture stress for each composition. Alloy A with zero percentage Nb has a good yield stress coinciding with maximum stress. Alloy B with 5 at% Nb content shows serration behavior (plastic deformation) of continuous drop and gain in stress after yield stress, which continues till a certain strain value before decrease in stress and failure. Alloy C (10 at% Nb) shows an appreciable increase in yield and maximum stress values but the fracture behavior is similar to alloy A without any serration and finally in the end, alloy D (with maximum 15 at% Nb) shows a dramatic decreases in ability to absorb stress before and after failure as compared to all other values. This is attributed to development of certain IMCs and other constituents at higher alloying element content, which might have caused decrease in stress.

1.15 VERY RECENT TRENDS AND TRIUMPHS

Some of the modern approaches to the problem of achieving ductility and toughness are fundamental in nature based on basic understanding and comprehension of engineering and metallurgy. For example, a recent study details the size effects on stability of shear band development and propagation. This interesting review documents very recent developments and progresses in ductile BMGMCs in the form of important phenomena of shear banding, which ultimately results in increased ductility and toughness in otherwise brittle solids (Yang and Liu 2012). As discussed in Section 1.11, formation of stress-induced transformation or transformation induced plasticity (TRIP) inside a ductile phase dendrite is another promising way of achieving large ductility while maintaining high strength and hardness. Although it is relatively old idea, which was exploited some years ago by means of indentation and conventional deformations (Jiang et al. 2005; Flores 2006; Jiang and Atzmon 2006; Shi and Falk 2007; Dodd and Bai 2012), it has attracted the attention of researchers as new methods of forming and transformation (especially since *in situ* liquid–solid transformation [Song et al. 2016]) have evolved with time. The quest for obtaining a ductile BMG composite with enhanced optimal ductility with large enough size still continues to push boundaries of what could be achieved. In this regard, very recently, researchers at Yale University and IFW, Dresden, Germany have made further promising progress, the details of which could be found in Sarac and Schroers (2013).

1.16 LIMITATIONS/RESEARCH GAP

Despite advances and triumphs, there are still a number of unresolved issues from processing (chemistry, physics, metallurgy, and engineering [tooling and machinery]), structural (phase identification and their behavior), properties (mechanical, physical, and functional) viewpoint, which limits their application and further use in more advanced applications, commercialization, and large-scale production. For example, despite being able to be produced in bulk form, still the largest ingot casted as BMGMC is just 80 mm in diameter and 85 mm in length (Nishiyama et al. 2012). Liquid metal technologies have been able to produce various types of shapes in "cast" form but these are by adopting very expensive tooling and are very thin in their profiles (Schroers 2010). There are very few successful efforts to make parts with tensile strength greater than 980 MPa in Al-based BMGMCs (Akihisa et al. 1988). Despite its advantages, twin roll casting (TRC) remains a novice technique for fabrication of BMGMCs of all types. Only Ti-based BMGMCs could be produced with ease because of their increased fluidity. Zr-based BMGMCs still have biggest limitation for large-scale production as these are viscous and their transformations are sluggish because of suppressed kinetics. There is very little effort on the functional use of BMGMCs (Inoue 2001). Reproducibility of these composites is another outstanding debate and contradictions exist about their behavior from laboratory to laboratory. Effect of microstructural control parameters and its tuning with variety of materials and physical parameters is not known. Lastly, AM (Qian 2015; Sames et al. 2016), though promising technique and presently being named as "future," has serious drawbacks (microstructure, modeling, metallurgy, mechanical properties and anisotrophy) for the use of Al (Olakanmi et al. 2015), Fe (Kruth et al. 2004; Gibson et al. 2010a; Frazier 2014; Jung et al. 2015b), Ti (Gibson et al. 2010b; Dutta and Froes 2016), and Zr (Yue and Su 2008; Chen et al. 2010; Sun and Flores 2010, 2013; Yang et al. 2012; Zhang et al. 2015)-based BMGMCs.

1.17 PRESENT RESEARCH—BRIDGING THE GAP

In the present research, an effort has been made to microstructurally control and tune the properties of Zr-based BMGMCs by controlling the number density (d_c) of ductile second phase (B2), its grain size, and dispersion in bulk alloy by conventional and AM routes. This novel idea stems from the fact that the inoculation of otherwise passive melt can cause precipitation of certain phases prior to other microstructures in an alloy. This can effectively be used for evolution of preferred phases, thus affecting properties. It is envisaged that careful selection of potent inoculants which can best serve as sites for preferential nucleation of ductile phase only can best be used to increase their number density and dispersion in bulk of alloy. It has been previously reported that 3D arrangement of network of ductile phase equiaxed dendrites in bulk alloy can effectively serve as source of impediment of shear band motion and can best serve as a junction for their multiplication (Hays et al. 2000, 2001). Further, there are methods by which only high-potency inoculants whose crystal structure matches the crystal structure of precipitating phase can be preferentially selected as compared to other inoculants. This is known as "edge to edge matching" (Kelly and Zhang 1999, 2006; Zhang and Kelly 2005a, 2005b). Selection of nuclei by this method and then controlled inoculation by them can serve as an effective means for increasing the number density, size, and distribution of ductile phase dendrites in bulk. This fact is successfully exploited in present research. During the course of study, computational model based on probabilistic cellular automation was proposed to be developed which will be used to predict the size, shape, and morphology of dendrites and their evolution. The model takes into account the effect of crystallographic orientation and motion of liquid–solid front as well. This is proposed to be coupled with a transient heat transfer model in the melt pool of additively manufactured part (laser materials interaction region). A code of model is aimed to be developed in MatLab Simulink and its coupling is proposed to be done by SolidWorks and Ansys. The results predicted by computational studies are proposed to be verified by their observation in actual fabricated samples in SLM machine. This experimental verification is proposed to be done by optical and electron microscopic analysis.

SECTION 2

ADDITIVE MANUFACTURING

2.1 WHY ADDITIVE MANUFACTURING?

Additive manufacturing (AM) of metal parts has recently evolved as the most important, versatile, and most powerful technique bearing a lot of futuristic potential of full scalability to develop complete parts in one run without any restriction of size, shape, material(s), and complexity. In addition, due to its inherent nature (laser [Pinkerton 2016] or high-energy [electron] beam [Romano et al. 2015] assisted sintering (SLS) or melting (SLM), layer-by-layer (LBL) formation, and other associated processes (denudation zones, Marangoni convection, and plasma), it eliminates the need of any posttreatment (heat treatment, trimming, grinding, finishing, and polishing). In most cases, it is direct production of actual (to be serviced) part in full, just in one step. Minor trimming, shaking, or air blasting may be needed, which does not comprise the bulk process. These unique features have raised its importance and attracted a lot of attention from the scientific, technical, and engineering community to further promote its use, work toward its progress, and remove bottlenecks in its operation.

2.2 A BRIEF HISTORY OF ADDITIVE MANUFACTURING

AM has been in use since long in various forms which primarily consist of shaping of metals, alloys (Fe based [Kruth et al. 2004; Rombouts 2006] including steels [Badrossamay and Childs 2007], Al [Buchbinder et al. 2011; Kempen et al. 2012], Mg [Taltavull et al. 2012], Ti [Vandenbroucke and Kruth 2007; Warnke et al. 2008; Leuders et al. 2013], and Cu [Tang et al. 2003] bases), ceramics (Shishkovsky et al. 2007; Yves-Christian et al. 2010), glasses (Quadrini and Santo 2008; Luo et al. 2014), and polymers (Berzins et al. 1996; Kandis and Bergman 1999; Childs et al. 2000; Mazzoli et al. 2007; Shinno and Yamada 2009) into different forms by the use of highly focused, localized, and concentrated source of energy. Traditionally, after the discovery of electricity, for metallic parts, this energy has been supplied in the form of electric arc which melts the base metal, thus providing a small consistent melt pool which is maintained by the help of external addition of material(s) (powder, fluxed wire [filler], etc.) and serves as the source of joining, deposition (coating, facing, and cladding), or fabrication (repair). Modern-day AM evolved from its predecessor which was primarily laser-assisted coating of materials (hard facing, thermal barrier coating, and biomaterials coating) on other materials. It evolved into "cladding" (Yoshioka et al. 1987; Wu and Hong 2001; Wu et al. 2002; Yue et al. 2007; Zhu et al. 2007b; Yue and Su 2008; Hui et al. 2010; Schmidt et al. 2011; Zhang et al. 2011a, 2011b; Zhang et al. 2011a, 2011c; Wu et al. 2015b; Harooni et al. 2016), which was a form of welding involving continuous or semi-continuous deposition of one material on top of other to make a thick layer which serve as protective (corrosion, wear, heat, etc.) or face layer. The latest predecessor of AM is laser (rather than electric arc) assisted "gas metal arc welding" or "metal insert gas" welding (Li et al. 2006; Kim et al. 2007; Kawahito et al. 2008; Wang et al. 2010b, 2011, 2012a) using various types (fiber [Kawahito et al. 2008], Nd:YAG [Glardon et al. 2001; Kim et al. 2006, 2007; Wang et al. 2010, 2010b]) and operating modes (continuous and pulsed [Lin et al. 2012; Yang et al. 2012] micro, pico [Quintana et al. 2009], and femtosecond [Wang et al. 2007; Jia et al. 2008; Ramil et al. 2009; Ma et al. 2010; Yang et al. 2012; Miracle et al. 2014]) of lasers. This setup is very close to modern-day AM setup. Only difference being, in AM, part is produced in LBL fashion from a sliced CAD pattern fed to a computer controlling the setup at the back end. The technique matured with time to make full-scale components from various types of powder feed stocks and base metals as it attracted the attention of various groups around the globe (Sun and Flores 2010; Yang et al. 2012; Li et al. 2014b; Brif et al. 2015; Olakanmi et al. 2015; Tang et al. 2015; Zhang et al. 2015; Borkar et al. 2016; Körner 2016).

2.3 PRINCIPLES OF ADDITIVE MANUFACTURING

Although present in, and known by, different variants (direct laser fabrication, direct metal deposition [DMD] [Marion et al. 2015; Amine et al. 2014b; Marion et al. 2015; Sames et al. 2016], laser metal deposition [Shukla and Verma; Zekovic et al. 2005], direct laser deposition [DLD] [Amine et al. 2014a; Joseph et al. 2015; Shamsaei et al. 2015; Thompson et al. 2015], direct form fabrication, direct digital manufacturing, rapid prototyping [Shiomi et al. 1999; Santos et al. 2006], 3D printing, powder feed method (PFM) [Liu and Qi 2014], wire feed method [WFM] [Ding et al. 2015], selective laser sintering [SLS] (Kathuria 1999; Kumar 2003; King et al. 2015; Hebert 2016), selective laser melting [SLM] [Lee and Zhang 2016; Matsumoto et al. 2002; Kruth et al. 2004; Sun and Flores 2010; Thijs et al. 2010; Lee 2015; Olakanmi et al. 2015; Prashanth et al. 2015; Spears and Gold 2016], laser engineering net shaping [LENS] [Matthews et al. 2009; Griffith et al. 2011]), basic principles of AM remain the same (Gibson et al. 2010a; Pinkerton 2016). It consists of energy source (laser [Pinkerton 2016] or high-energy electron beam [Romano et al. 2015; Körner 2016]), powder (serving as bed beneath, fed alongside [coaxial], or at an angle to energy source), a drawing/pattern generation system (CAD computer), slicing system (slicing software), and a mechanical ratchet-based or mechatronics-controlled system for step by step downward motion of powder holding table. There are other auxiliaries of the system (such as roller, hoopers, alignment systems, switching, and viewing window) which in totality form the system but they are minor as compared to main parts described previously. Although developed with respect to its mechanical and mechatronics system, it still has its shortcomings, which require more research and development (R&D) specifically on materials properties and their final homogeneity across whole part (in case of very large parts and in thin sections), consistency of properties in very complex alloying compositions, and very little or no knowledge about melt pool, its shape, size, depth, dynamics, evolution, and transient behavior affecting final solidified metal/alloy amount and quality. This and other similar drawbacks are still challenges for its utilization, adoption, and full commercialization at large scale. These restrict its use primarily to repair and damage mitigation rather than full-scale neat part manufacturing. Below sections briefly highlight various major processes presently in use to form final high-quality parts.

2.3.1 LASER PROCESSES

Laser-assisted AM or simply laser-based AM processes are set of procedures in which source of energy for sintering or melting and subsequent solidification from powder bed or feedstock is provided by highly localized, concentrated, and focused laser light. These technologies can broadly be grouped into one of seven major classes based on the mechanism in which each layer is formed: photopolymerization, extrusion, sheet lamination, beam deposition, direct write and printing, powder bed binder jet printing, and powder bed fusion (PBF). For the purpose of the present study, these processes may be classified on the basis of how metal feedstock is made available, presented to, and interact with laser light. This can be powder bed processes, powder feed process, or wire feed process. Further, powder bed processes can be classified on the basis of how powder consolidation or fusion can take place. Thus, it forms the basis of SLS in the former case and SLM in the latter. A brief classification of laser-based metal AM processes is presented in Figure 2.1.

Every variant of AM has its own significance and applications. Their behavior can be classified into two major groups of variables: (a) production/machine variables and (b) process/materials variables. The former include type of machine, type of laser (CO_2, fiber), mode of operation (continuous or pulsed), mode of impingement on target, size of machine, scan depth, scan rate, and scan speed, while the latter may include type of target, its shape, size, geometry, form (solid, and powder), properties prior to AM, properties during AM, properties after AM, and its intrinsic and extrinsic properties. Both these fundamental types of variables and their optimization are very important in order to get the best properties and part quality out of AM process. Unfortunately, there is no one "rule of thumb" which can dictate the selection of a particular process best matched for an alloy. Due to noviceness and still infant nature of process, very little data is available on varied types of alloy systems which can prove one material's suitability to be best with a particular type of process. There has been generalized classifications (Kumar 2003; Sames et al. 2016; Smith et al. 2016) but they all are based on empirical and irrational data and ideas. The advent and use of advanced modeling and simulation technologies (Romano et al. 2015; Thompson et al. 2015; Khairallah et al. 2016; Vastola et al. 2016) to understand transient and incipient processes happening in AM have helped a lot in paving the way for the optimization of process as a whole and its suitability for a particular group of metals and alloys. However, the best optimized selection of a proper process which is best suited for an alloy or material is still at stone's throw and is evolving with time as more and more experimentation is being carried out.

2.3.1.1 Selective Laser Sintering

SLS, as the name suggests, is a laser-based AM process in which interaction of laser with powder metal bed causes localized sintering which is evolved in LBL fashion to finally fabricate a complete part which carries the impression of CAD drawing of component fed to machine via processor in preceding step. The process was invented in 1986 at the

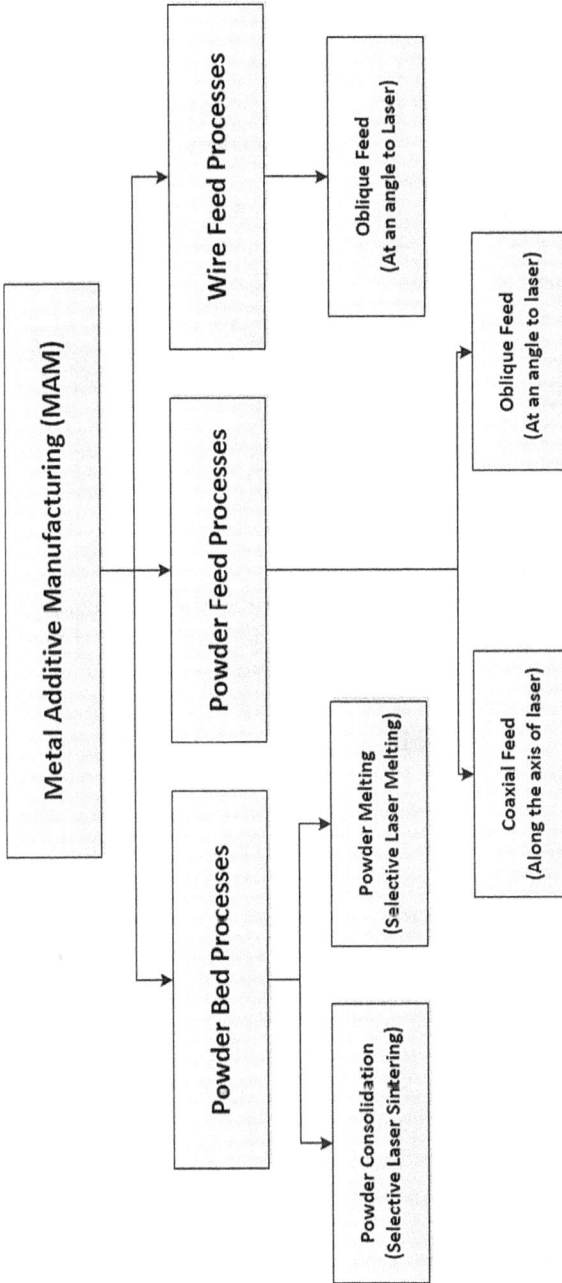

Figure 2.1. Classification of laser-based AM processes.

University of Texas when Dr. Carl Deckard was awarded the first patent for his invention (Deckard 1988). The process was first developed for the fabrication of 3D components (mainly prototypes for *audiovisual help* and *fit-to-form* test) from polymers and nylon. With the passage of time, it was extended to metals and alloys to manufacture *functional prototypes* and *rapid tooling*. The impulse in the process came from involvement of industrial clusters and computational techniques performed in laboratories to optimize the process parameters and enhance the efficiency of sintering machine (Kumar 2003). Overall aim in these processes is not to melt but heat the powder to a temperature below melting point in conjunction with other mechanisms (particle size, density, distribution, pressure, and binder) which promote their agglomeration. Based on mechanisms of how they get heated and microstructuring, SLS could be broadly subclassified into three classes (Kathuria 1999):

- Direct method: one component
- Indirect method: one component + polymer
- Two-component method

Direct method: one component: This is also called single-component solid state sintering. In this, laser energy is tuned to such an extent that it causes a temperature rise close to melting point of metal power but does not exceed it. This is done to cause binding at the interfacial grain contact area. This mechanism is called "particle fusion below melting point." This process involves formation of a neck at contacting surface of two adjacent particles which results in reduction of surface area, thus causing an increase in tendency of powder to aggregate. The driving force for this is reduction of free surface energy of particles and densification proportional to this reduction. The limitation with this process is precise control of metal temperature as well as particle size. For higher temperatures or complete melting, one finds that—owing to higher viscosity and surface tension effects—the molten metal tends to form a spherical ball-type structure which has dimensions larger than particle size. On top of it, as laser spot is usually lager than particle size, many particles get melted together and a bigger spherical droplet is formed. Because of large size of this spherical droplet, it is connected with other droplets only at certain points on its contour. Therefore, it is very likely that porosity will be formed in final sintered structure. This process is workable only at slow speeds but is difficult to carry out at high speeds encountered in typical industrial setup and usually postprocessing to eliminate porosity is required (not a desirable feature). *Indirect method: one component + polymer:* This method involves application of thin polymer coating on powder particles prior to laser sintering. This results in reduction of "balling effect" encountered in direct processes. In this, polymer behaves like a low-temperature phase serving as binder. Overall binder content is about 1% by weight. Due to its high infrared absorption and low melting point (~150°C), the polymer melts in advance to metal powder (low infrared absorption and high melting point [~1,488°C] and connects the metal particles together without causing melting. Final sintered parts are about 45% porous in green stage. The density of parts may be further increased by its posttreatment. The complete process (posttreatment) comprises of three stages and can be carried out in one furnace. It is namely (a) burnout, (b) high-temperature sintering, and (c) infiltration. During the first step, the temperature is raised to 300°C at which remaining polymer burns out followed by sintering at higher temperature (>700°C). This causes formation of porous skeleton of metal particles. Final densification may be achieved by infiltration with copper at 1,083°C at which Cu infiltrates the skeleton by capillary action. Final part is about 60% structural material and 40% Cu. Main *disadvantages* with the process are it is slow, time-consuming, and desired density may not be achieved as a result of shrinkage. *Two-component method:* This is a typical process used with two metals. One with higher temperature (T_2) called structural metal and the other with low-temperature (T_1) called binder. In this process, the laser energy is tuned in such a way that it reaches a temperature (T) between melting points of two metals: $T_1 < T < T_2$. This causes the binder metal to melt and flow under the action of forces such as liquid pressure, viscosity, and capillary action through the pores between the solid particles. Therefore, pores are reduced and strength of structure remains intact.

Case 1: If $d_1 < d_2$
For an enhanced densification, the grain size of binding particles (d_1) (Metal 1 with T_1) should be smaller than that of structural material (d_2). The reason for this is the difference in melting enthalpy of particles of different sizes. Bigger particles have higher melting enthalpy due to their large size, becomes less susceptible to melting, and therefore are well wetted. On the other hand, particles of small size have lower melting enthalpy, lose their structure completely, and form a cluster structure. Thus, larger particles are well wetted by smaller particles.

$$\text{Particle size} \propto \text{melting enthalpy} \qquad (1)$$

Case 2: If $d_1 > d_2$
Open porous structure will be formed. This could also happen due to lower laser energy causing an effect called "residual porosity." On the other hand, if laser power is too high, excess liquid formation may produce compact

distortion (not desirable). Therefore, (a) an optimum laser power (energy) and (b) aspect ratio of binding material to structural material are very important parameters for this process. Over the course of whole process (total time for sintering), three distinct stages may be identified. *Stage 1*: During this stage, melting, wetting, or liquid flow, rearrangement and densification happens. *Stage 2*: This is dominated by pore removal and solute participation. In this, capillary forces compress the particles together resulting in volume shrinkage. *Stage 3*: In this last stage of solution precipitation, interconnected pores pitch to form isolated pores.

Overall, the first stage is dominated by rearrangement, second is by shape accommodation while filling of pores and grain coarsening happens in final third stage. In laser sintering, usually a thickness of a few hundred μm/layer is achieved. The operating details of process may be found elsewhere (Kumar 2003; Kruth et al. 2004; Olakanmi et al. 2015). It is rarely used for manufacturing of bulk metallic glass (BMG) and their composites (BMGMC) because of their multicomponent nature and binder issues.

2.3.1.2 Selective Laser Melting/Laser Engineered Net Shaping

This is a process which is distinctly identified by complete melting of a region of powder bed at which laser–matter interaction occurs, followed by maintaining of this liquid melt pool and its traversing all through the shape of part to be produced at all times of AM. This distinctively gives it the name SLM. From a conception point of view, it is very much similar to SLS described in the previous section. It consists of 3D drawing (usually in CAD) of part to be produced. This 3D drawing is sliced using a slicer (a software) usually in layers of 20 to 100 μm thick. This creates 2D image of each layer. This file is in standard stereolithography (.stl) format accepted by industry. After this, file is imported into another file preparation software package (machine integrated) which assigns parameters, values and physical supports to the file such that it could be interpreted by different industrial additive manufacturing machines. However, from mechanism point of view, it is completely different that it involves complete micro–macro scale melting (fusion) followed by holding and then solidification. This results in solidification microstructure of metal or alloy to be deposited which follows standard metallurgical principles (phase diagrams, liquid–solid transformations, solute partitioning, and crystallography) for its progression and formation. History of SLM goes back to 1995 when Dr. Wilhelm Meiners and Dr. Konrad Wissenbach of Fraunhofer Institute ILT in Aachen, Germany were awarded the first ILT

SLM German Patent DE 19649865 (Meiners, Wissenbach, and Gasser 1998). Since then these researchers are the key names in further development, progression, and advancement of this technology from different platforms founding numerous companies and industrial clusters. Detailed mechanical (laser metal interaction [heat generation and transfer]) and metallurgical mechanisms (alloy solidification, and phase transformation) are subject part of this study and will be described in detail in later sections. This section merely concentrates onto introducing the basics of SLM from process and operation viewpoint.

A typical SLM process is illustrated in Figure 1.14. It consists of "powder hopper" or dispenser bed which carries the load of powder. An elevation mechanism below powder reservoir lifts a prescribed amount of powder above the level of build plate which is then spread in thin even layer over the build surface by a relocator mechanism (scraper, roller, or soft squeegee). Another variant (as highlighted) may supply powder from above the build platen (more easy configuration and used in American machines). The thickness of powder layer is typically 10 and 100 μm. A high energy source (laser) is made to impinge on the surface of powder bed which causes its melting due to heat generated. These lasers could be of various types (as described earlier) but normally they are fiber lasers with wavelength 1.06 to 1.06 μm and power on the orders of hundreds of Watts. After the first layer (which is melted) is solidified as laser traverses the complete scan over whole part dimensions fed in the form of CAD file, the same elevator mechanism lifts the powder bed or hopper supply powder from top again to make a thickness of few microns, scraper or roller smoothe out the powder layer, and the process is repeated again. This results in LBL fashion buildup of part conforming to dimensions and geometry fed at earlier stages (CAD level). The process is usually carried out under protective (Ar) atmosphere to avoid oxidation and contamination. There are over 50 process variables which need to be understood and controlled to ensure they achieve a good quality part. The details of all would be exhaustive. However, they can be generally classified into four categories: (1) laser and scanning parameters, (2) powder material properties, (3) powder bed properties and recoat parameters, and (4) build environment parameters. These can be further classified into controllable parameters that can be manipulated during a build process and predefined parameters that are determined at the start of a build and remain essentially set throughout the process. A detailed treatment of these could be found in excellent reviews (Van Elsen 2007; Mani et al. 2015; Spears and Gold 2016). LENS (Figure 2.2) is a special variant of melting/fusion processes in which powder is preferentially fed from overhead hoppers coaxially with laser in such a way that

Figure 2.2. Laser engineered net shaping (LENS) of turbine blade.

it goes directly to melt pool and motion of "combined laser–powder feed head" keeps on traversing its path dictated by CAD geometry at the back end which finally results in 3D shape. The process was developed at Sandia National Laboratories in 1996 when Jeantette et al were awarded first patent (US Patent 6,046,426) and drives its name from there (Jeantette et al. 2000). The details about process description (Griffith et al. 1996; Atwood et al. 1998), process optimization for surface finish, and tuning of microstructural properties (Smugeresky et al. 1997) are described by original inventors in articles thereafter. Owing to excellent configuration, procedure of how powder is introduced into melt pool, and how it interacts with laser, the process has found excellent applications in new variant methods. These include functionally graded materials (Tejaswini et al. 2015) and manufacturing of parts with spectrum of properties across their cross section which can be tailored to arrive at particular property at a specific location.

2.3.1.3 Direct Energy Deposition,Powder Feed Method PFM, and Wire Feed Method WFM

DMD, DLD, or powder feed method is a form of direct energy deposition (DED) process in which a concentrated source of heat (laser/electron beam [explained in the next section]) is made to impinge on a surface which is utilized for melting along with *in situ* delivery of feed material (powder/wire) for LBL fabrication of complete part or single-to-multilayer cladding/repair. The process differs from LENS (which is a form of DED) in that the delivery of feed material could be coaxial or oblique (at an angle) to laser/electron beam direction. The process has evolved as major competitive method for (i) rapid prototyping of metal parts, (ii) production of complex geometries, (iii) cladding/repair of precious metallic components, and (iv) manufacture/repair in recesses and difficult to approach regions. Like other processes, it also involves thermal/fluidic and transport phenomena which occurs in different regions:

a. Laser–metal interaction (heat generation)
b. Development of melt pool (fusion)
c. Shape, size, configuration, and traversing of melt pool (heat and mass transfer)
d. Metal deposition (solidification—nucleation and growth (NG) and part scale heat transfer) and
e. Microstructural evolution (solute diffusion and capillary action effects)

History of process again dates back to same period of mid-1980s and 1990s when other processes (variants of DED) emerge on global surface. Principally, they all consist of laser–metal interaction in one way or other, which is why their fundamental principle does not change and remains the same. Machine setup in these processes is relatively simple. It consists of built-in software which automatically checks most sensors. Powder hoppers are filled and a build substrate is positioned below them on a stationary (three-axis systems) or a rotating (5+ axis systems) stage. This increases the ability of machine to produce more complicated shapes (Sames et al. 2016). Like PBF processes, DLD process also comprises of various operating/process parameters. Setting, monitoring, and controlling of these parameters are dictating pivot to ensure process and part quality to a large extent. Again, these could be categorized into two main types: (a) machine/process parameters and (b) material parameters. Most important parameters of these are

• *Laser/substrate relative velocity (traverse speed):* This dictates the length of time taken by a DLD to build certain geometry. It usually is on the order of 1 to 20 mm/s.

- *Laser scanning pattern:* It dictates the laser position and height-wise positioning of the substrate via numerical control set by an operator.
- *Laser power:* This is the total emitted power (in Watts) from the laser source. It is typically on the order of 100 to 5,000 W.
- *Laser beam diameter:* These are usually on the order of 1 mm.
- Hatch spacing
- *Particle/powder feed rate:* This is the average mass of particles leaving the DLD nozzle per unit time (typically 1 to 10 g/min). This usually should be continuously monitored and nozzle cleaning and other maintenance be performed if this rate is impeded.
- *Interlayer idle time:* The interlayer idle time is the finite time elapsed between successive material/energy deposits.

Generally speaking, these operating parameters are *material dependent* and vary in conjunction with (a) *DLD machines (e.g., number of nozzles, nozzle design)* and (b) *operating environment. Powders* can vary in (i) size, (ii) shape, and (iii) their production method. For most laser deposition processes, powders are larger in size as compared to those used in fusion processes (as there is no constraint [distance] between them and source of heat [laser]). These typically are 10 to 100 μm and are spherical in shape. Spherical-shaped particles have the advantage that they retard entrapment of gas in melt pool (as they are introduced in it) and thus results in final part with little or no porosity. For production of alloys, typically gas or water atomized or plasma rotating electrode processed powders are used (Qian 2015). Build chamber in case of DLD is enclosed to provide laser safety; however, it is not a compulsion to fill it completely with insert gas to provide shielding (as is done in case of SLS/SLM processes). For nonreactive metals, shield gas directed at the melt pool suffices while for reactive metals (Ti, and Nb), chamber is flooded with an inert gas (Ar or N_2) after necessary vacuum to provide additional shielding and safety (Sames et al. 2016). Detailed theoretical (Thompson et al. 2015) and experimental (Shamsaei et al. 2015) description of process is given in two very recent articles by Prof. Thompson and his group at Mississippi State University.

2.3.2 ELECTRON BEAM MELTING

Electron beam melting (EBM) is an AM process in which source of energy for melting is high-energy, localized and focused electron beam rather than laser. The main difference with EBM is that it can only be applied to metallic components since electrical conductivity is required. However, the advantages are that it can be moved at extremely high velocities (scan speed) and beam power is merely a function of available power from grid. Materials properties initially were adversely affected by amount of porosity generated during the process which now has been overcome to a large extent. Nowadays, treatments by highly focused and concentrated electron beam result in very finely dense structure whose properties are comparable to properties achievable by other processes such as casting and welding. In addition, there is high potential of the process—inherent, rapid, and directed solidification which leads to very fine microstructure and epitaxial growth (a unique feature only possible in EBM) (Körner 2016). History of AM goes back to the time when electron beam was harnessed to be used for imaging (microscopy) and subsequently some studies were reported in United States in the mid-1980s (Tauqir 1986). They were not carried out with respect to using electron energy for part buildup. The objective was only to carry out melting using electron beam. Real progress in this field came in 1997 when Arcam AB Corporation in Sweden commercialized the first selective electron beam melting (SEBM) machine (Sigl et al. 2006; Kahnert et al. 2007; Tang et al. 2015). The process is fundamentally very similar to scanning electron microscopy (SEM) which involves generation of an electron beam by heated W filament, their collimation and acceleration to 60 keV (usually much higher than that in microscopy), and finally impingement on build table housing powder after passing through two magnetic focusing coil (magnetic lens) system. Beam current is controlled in the range 1 to 50 mA and the beam diameter (Ø) ~0.1 mm. Powder size typically ranges from 10 to 100 µm and typical layer thickness it forms on the bed is 0.05 to 0.2 mm. Electron beam scans the surface in two passes. *First pass* consists of traversing path at higher speed (~10 m/s). This step is repeated multiple times and is carried out to preheat the powder to sintered state. This is followed by second slower scan (~0.5 m/s) during which melting occurs. Once the top layer is solidified, a roller lays down another layer of powder and the process is repeated. The entire process takes place under vacuum (typical 10^{-1} Pa in vacuum chamber and 10^{-3} Pa in electron gun). In addition to vacuum, a low-pressure inert gas helium is added to the chamber at a pressure of 10^{-1} Pa to avoid buildup of charge of powder. Once part built-up is complete, part is allowed to cool inside chamber which may be assisted by further purging of He (Gong et al. 2012). Generally speaking, the path electrons traverse to build up a surface can be categorized into three roasters, namely heating, melting: hatching, and melting: quasi multibeam. These are briefly described in Figure 2.3.

The process is virtually suited to all types of metals and alloys resulting in various types of metallurgy (Murr 2015), melt pool geometry, its configuration, temperature distribution in it (Romano

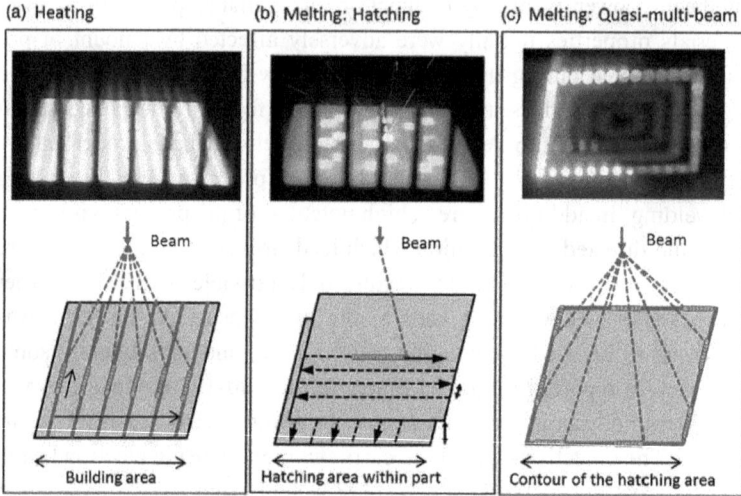

Figure 2.3. Actual pictorial (top) and schematic (bottom) representation of heating and melting during SEBM (Körner 2016).

et al. 2015), and type of materials handled and processed (Tang et al. 2015; Körner 2016). The process has been recently studied by the help of computational modeling and simulation for thermal heat transfer (Vutova and Donchev 2013) and microstructural evolution (Gong and Chou 2015; Vastola et al. 2016). There have been few studies aiming at EBM of glass-forming alloys in early (Bergmann and Mordike 1981) and late 1980s (Bhanumurthy et al. 1988), but they were primarily aimed at carrying out melting using electron energy rather than part buildup. Recently, few reports have been published about production of BMG by use of EBM (Koptyug et al. 2013; Drescher and Seitz 2015) out of which notable effort has been made at Mittuniversitetet, Sweden in collaboration with Exmet AB, but it is the first of its kind and a lot of R&D effort is needed to successfully produce large and complex BMGMC parts using EBM.

2.4 CHARACTERISTICS OF PROCESS

There are many unique characteristics of the AM process which give it a distinct position and advantage over other manufacturing processes. Some of these include LBL formation, rapid cooling, and *in situ* heat treatment of base layer. These will be briefly described here.

2.4.1 LAYER-BY-LAYER FORMATION

LBL formation is a unique feature of AM. It is a name given to evolution of unique solidified layers which are formed on top of each other as the process continues. It gives rise to layered structured deposition of metal/alloy in built-up fashion. Usually, it is associated with two parameters: (a) vertical build direction and (b) cross-build direction (Figure 2.4).

Both have their own unique features/role in AM and effect on final alloy quality. Vertical build direction is associated with building of alloy in a direction normal to plane of rest or axis of table (X-axis). This involves deposition of metal layers on top of each other in such a way that finally a part is built which carries mass which is accumulation of welded/joined layers. As described in the previous section, this built is highly a function of scan speed, laser energy, scan depth, and material chemistry. The resultant part, its final chemical makeup, and compositional homogeneity are dependent on how accurately and precisely aforementioned parameters are controlled. Excessive time spent at one point will result in localization of heat and may cause extreme deterioration, which may include burning of part. This is very critical in low-melting (Al- and Mg-based) alloys. In high-melting metals (single-component or two-component [binary] alloys), this is not a big problem but in multicomponent alloys (BMGs, HEAs) this may cause loss of chemical composition due to burnout (oxidation) of low-melting alloying elements. Other detrimental effects of this are: segregation, anisotropy, crystallographic distortion (elongation of crystal axis), improper and incomplete melting and fusion between layers, and formation of unwanted phases and intermetallics (IMCs), which can severely hamper the final alloy's mechanical properties. Usually, optimization in this direction is highly a function of operator skill and experience. However, a lot of help is provided in modern days by data and visual stimulated effects and graphics produced by modeling and simulation of various concurrently happening phenomena (Gusarov and Smurov 2010; Khairallah et al. 2016) during AM. Buildup along X-axis (i.e., along axis perpendicular to normal built direction) is another important parameter which must be controlled properly, adjusted, and optimized to get good quality part. Any improper functionality in this will result in improper fusion between adjacent layers, which may give rise to porosity and undefined porous structure which not only weakens the final alloy but is also the source of bad appearance and make it prone to atmospheric and weather (corrosion) attacks. On the other hand, use of very high laser power while building LBL pattern in adjacent layer will result in a fusion, melting, and compositional variation at points of contact (tack points) between LBL of first and second layer. This is a very good example of tack to tack failure which is very common under fatigue loading.

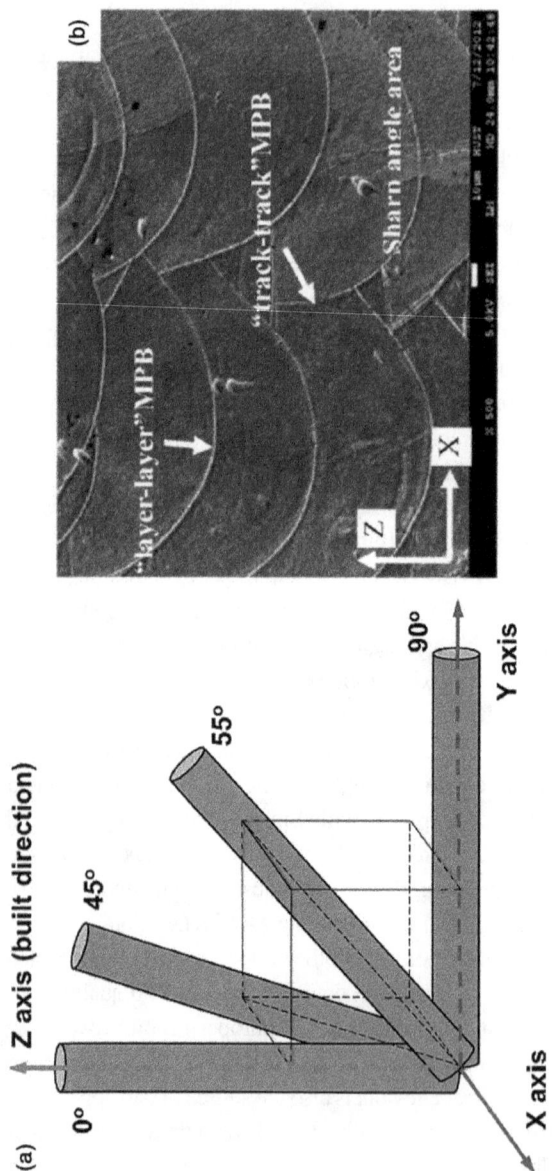

Figure 2.4. (a) Schematic of build directions (Sun et al. 2014), (b) typical micrograph showing vertical build direction (Y), cross-build direction (X) and their effects (Shifeng et al. 2014).

2.4.2 RAPID COOLING

The second most important feature of AM is rapid cooling. It results from very shallow depth and very narrow size (length and width) of melt pool. In addition to this, it has transient nature. That is, it keeps on changing its shape and pattern of heat transfer from it during the time laser scans its path following CAD geometry. These features of melt pool make it highly unstable and it tends to quickly return to its stable form by dissipating heat away from it (solidifying). This gives rise to rapid cooling. This rapid cooling is particularly beneficial in multicomponent alloys where it facilitates texture development and directionality of properties. For example, in case of additively manufactured turbine blades, their properties are comparable (in fact superior) to directionally solidified blades by conventional means. This rapid cooling also reduces the fusion zone and thus decreases the chances of spatter, heat-affected zone (HAZ), fuming, and other detrimental welding effects. In case of BMGMCs it is highly desirable as it helps in *in situ* glass formation which is highly desirable with recourse to an additional processing step and without adding to additional processing cost.

2.4.3 IN SITU *HEAT TREATMENT*

Another very big feature of AM is *in situ* heat treatment of layer beneath fusion layer. As the primary layer which is raised to melting temperature (T_m) by tuning of laser power to high value, it traverses its path following CAD geometry. It heats the layer beneath it which is already solidified. This heating is sufficiently high enough to cross first transformation temperature (lower critical temperature/invariant temperature) in most alloys. This triggers the first solid state transformation (solid–solid transformation) which usually is precipitation out of stable crystal out of metastable highly (nonequilibrium) cooled products in first layer. This *in situ* heat treatment is integral part of most AM processes (at least those involving fusion [SLM/LENS]). In addition to formation of stable crystal, it helps in homogenization, normalizing, improvement in mechanical properties, and in some cases grain refinement of final alloy structure. In BMGMCs, this causes "devitrification" which is highly beneficial to increase toughness and ductility of this class of alloys. A noticeable disadvantage of this heat treatment is formation of intermetallic compounds (IMCs) which may form as a by-product of solid–solid transformation reactions. In the worst case scenario, incipient fusion/melting will occur at a point where low melting point constituents exist resulting from introduction/inclusion of impurities from first high temperature layer formation process.

This not only can mechanically weaken the alloy but also cause formation of porous network which is an additional impulse to decrease its quality. An effective way to avoid this heat treatment is to control machine parameters (laser power, scan rate, spot size) in such a way that only necessary amount of heat is transferred from laser to metal without causing development of zones of heating.

2.5 BULK METALLIC GLASS MATRIX COMPOSITES BY ADDITIVE MANUFACTURING

Processing of BMGMC by AM (Schroers 2010; Pauly et al. 2013) is slowly, progressively but surely growing as successful technique for their production on a large scale. Various forms of AM processes (SLS, SLM/LENS [Smugeresky et al. 1997], DLD [Shamsaei et al. 2015; Thompson et al. 2015], EBM) are slowly but surely attracting the attention of scientists around the globe to exploit their potential to be used as large-scale industrial technique(s) for the production of BMGs. Despite the inherent bottlenecks in the AM processes, there have been successful reports about their production preferentially by SLM—a form of AM involving complete fusion. Various types of glassy structures, for example, Al (Li et al. 2014b, 2015), Zr (Yue et al. 2007; Sun and Flores 2008, 2010, 2013; Yue and Su 2008; Chen et al. 2010; Wang et al. 2010; Yang et al. 2012; Zhang et al. 2015; Harooni et al. 2016), Fe (Zheng et al. 2009; Jung et al. 2015b), Ti (Vandenbroucke and Kruth 2007), and Cu (Sun and Flores 2008)-based BMGMCs have been successfully produced using SLM (AM).

It is well known that incipient metal fusion, its transience, progression (movement), and subsequent deposition out of melt pool following metallurgical principles (solute partitioning, alloy diffusion, and capillary action to form dendrites) follow a layer by layer (LBL) pattern. This LBL is distinct in almost all AM processes (SLS/SLM, DLD, EBM). The unique feature of this LBL pattern is that the top layer gets heated to metal/alloy temperature courtesy of tuning of laser power. This is called "fusion layer." Now, as the fusion layer traverses its path dictated by CAD geometry fed at back end (.stl file), it generates a Heat Affected Zone (HAZ) preceding its tip. This HAZ is very much similar to HAZ observed in other fusion welding processes. The metal following it is usually found in solidified fine equiaxed grain form. This tendency is good normal behavior of the fusion layer and results in good glassy structure (high GFA) in BMGs provided melt pool temperature is high enough to cause complete melting and heat is rapidly quenched out of it making a monolithic glassy structure. This results in hard brittle layer. Now, as the complete path in this layer is traversed, it is descended by few microns (dictated by initial alloy properties and machine parameters) and is supplied with new layer of metal/alloy powder by the help of scraper/roller. The laser again starts traversing the path fed to it in the form of sliced CAD pattern. This layer again reaches melting temperature and incipient fusion/melting takes place at laser/metal contact point. However, this time, a unique new phenomenon takes place. As the layer currently in contact with laser melts, it generates enough heat for the layer beneath it to reach a certain high temperature as well (usually 0.5 T_m and $> T_x$). This heating of lower layer is enough to take the alloy back into nose region of TTT diagram which causes its crystallization (sold–solid transformation (devitrification)). Depending on the alloy chemistry and amount of time spent at temperatures above T_x (in nose region of curve), there could be (i) complete glassy structure, (ii) partial glassy structure, or (iii) complete crystalline structure (no glass). The last is usually meant to be avoided during BMGMC processing and second is desirable.

There is, however, a very narrow window of composition and temperature during which complete glass formation or complete crystalline structure formation could be avoided. (a) Only alloys with very high glass-forming ability (GFA) should be selected from composition perspective and (b) should be tailored to cool with sufficient enough cooling rate (calculable from exact TTT diagram) which should cause their *in situ* equiaxed ductile phase dendrite formation during primary solidification in first layer retarding complete glassy state or crystallinity. Once *in situ* structure is formed, reheating of lower layer to temperature in nose region of TTT diagram during devitrification does not have much effect on further crystallization (due to kinetics [solute partitioning]) provided it should not be purposefully allowed to stay there for long. In general process, from a fundamental theoretical standpoint, 100% monolithic glassy structure or glassy matrix with fully grown *in situ* crystalline dendrites do not further undergo transformation to another crystalline phase (as they have already come out of their metastable glassy state). A powerful impulse on this could be caused by the introduction of carefully selected potent inoculants which are added to alloy melt during melting stage. These may serve as active nuclei for the preferential heterogeneous nucleation of ductile phase dendrites during primary solidification ensuring the least formation of metastable glassy state which in turn reduces the possibility of conversion of glass to crystallites during subsequent heating of layer (devitrification stage) as there is no glass (all the metastable or unstable phase have already been transformed to their thermodynamically stable state). No such effort has been made in the past to exploit this unique crystallographic feature of alloying in AM. This forms the basis of the present research.

Few leading groups in the world have recently produced BMGMCs by AM. A brief tale of some of these is narrated here. Sun and Flores (2010, 2013) successfully studied the effect of heat input on microstructure of Zr-based BMGs manufactured via LENS. They observed the formation of unique spherulites within HAZ at high laser input (10^4 K/sec) which disappeared as laser power is reduced (Figure 2.5).

These spherulites bearing unique crystal morphology seem to bypass isothermal cooling microstructures—a phenomenon not observed previously. The same effect was observed in their earlier studies on Cu-based BMGs (Sun and Flores 2008). In another study, Dr. Mark Gibson with coworkers (Welk et al. 2016) studied the effect of compositionally gradient alloy systems to manufacture BMGs and HEA composite layers via LENS. They aimed at finding an optimized composition at which effect of both alloy systems can be obtained in conjunction. Alloy systems consisting of $Zr_{57}Ti_5Al_{10}Cu_{20}Ni_8$ (BMG) to $CoCrFeNiCu_{0.5}$ (HEA) (first gradient) and TiZrCuNb (BMG) to $(TiZrCuNb)_{65}Ni_{35}$ (HEA) (second gradient) were used and processed at 400 W, 166 mm/s, and 325 W, 21 and 83 mm/s respectively. Using selected area electron diffraction patterns, they successfully reported the formation of fully amorphous region in the first gradient and amorphous matrix/crystalline dendrite composite structure (Figure 2.6) in the second gradient in individual melt pools.

Increasing the speed caused a slight variation in morphology and composition. Their results were consistent with their earlier investigations (Cunliffe et al. 2012; Welk et al. 2014). However, the effect of reduced power and/or increased speed is needed to validate GFA of these systems. Zhang et al. (2015), investigated the effect of laser melting in the form of surface remelting and solid forming on well-known $Zr_{55}Cu_{30}Al_{10}Ni_5$ hypoeutectic system. They observed that despite repeated melting of alloy four times on its surface (LSM) during single trace, there was no effect on its glassy state. However, during solid forming (LSF), distinct crystallization was observed in the HAZ between adjacent traces and subsequent layers after first two layers. A series of phase evolution was observed in as-deposited microstructure as it moves from molten pool to HAZ. In these microstructures, $NiZr_2$ type nanocrystals and equiaxed dendrites form from rapid solidification (liquid to solid transformation) during LSM while $Cu_{10}Zr_7$ type dendrites form as a result of crystallization of preexisted nuclei (solid to solid transformation) in already deposited amorphous substrate. This paved the way for better understanding and application of LSM and LSF in terms of GFA and crystallization. Another group at the University of Western Australia led by Prof. T. B Sercombe developed aluminum-based BMGs by SLM (Li et al. 2014b, 2014c, 2016).

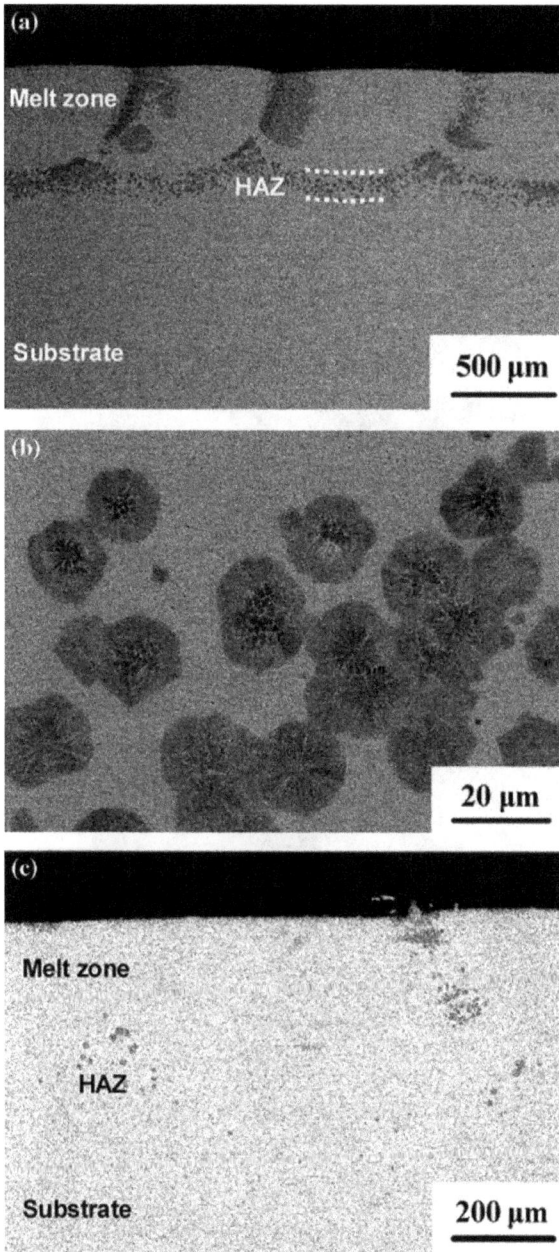

Figure 2.5. Cross-sectional backscattered SEM images of laser-deposited layers on the amorphous substrates processed at a laser power of 150 W. **(a) and (b)** microstructures obtained at a laser travel speed of 14.8 mm/s. The featureless melt zone is shown in (a) surrounded by a crystalline HAZ, and the isolated spherulites of the HAZ are shown in (b). **(c)** Increasing the laser travel speed to 21.2 mm/s reduced the formation of the HAZ to only a few isolated spherulites (Sun and Flores 2010).

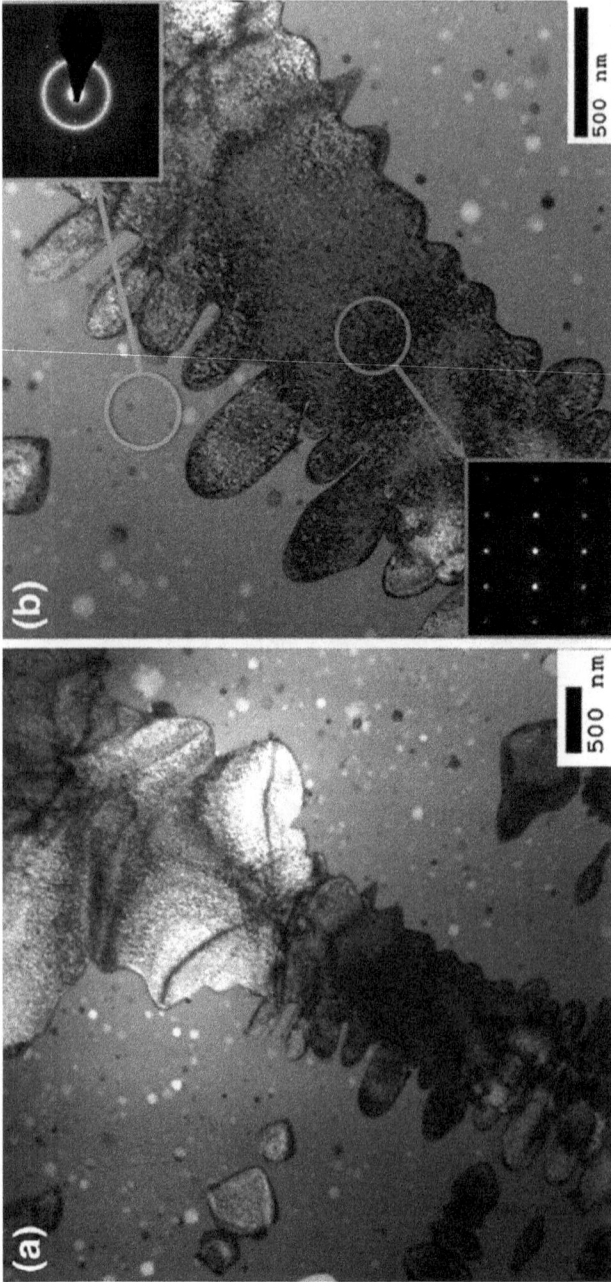

Figure 2.6. (a) TEM BF image of the laser surface melted region processed with a laser power of 325 W and a travel speed of 83 mm/s and (b) TEM BF image with the corresponding electron diffraction pattern of the crystalline dendrite (lower left inset) and amorphous matrix (upper right inset) (Welk et al. 2016).

They showed that an empirical laser power exists (120 W) at which width and smoothness of scan track are optimal, that is, defects (cracks [parallel, perpendicular, and at 45° to scan track] and pores) in scan edge are almost eliminated at this laser power. Crystallization, preferred orientation and melt pool depth are observed to have a direct relationship with laser power while pool width is observed in inverse relationship. Four distinct regions of Scan track (fully crystalline [~100 nm], partially crystalline [~500 nm], boundary between amorphous BMG and bigger crystals and edge of HAZ [no crystal]) are identified. They further studied preferred orientation and found it to be a major effect of devitrification (both by very high laser power [pressure wave] and temperature [oxidation]) as measured by EDS.

A few more notable studies have been reported very recently by leading research groups around the globe in which $Fe_{68.3}C_{6.9}Si_{2.5}B_{6.7}P_{8.7}$ $Cr_{2.3}Mo_{2.5}Al_{2.1}$ (at.%) (Jung et al. 2015b), Fe-Cr-Mo-W-C-Mn-Si-B (Balla and Bandyopadhyay 2010), other Fe-based BMGs (Basu et al. 2008; Matthews et al. 2009; Zheng et al. 2009), Ti–24Nb–4Zr–8Sn (Zhang et al. 2011b), other Ti based BMGs (Dutta and Froes 2016), $Al_{85}Nd_8Ni_5Co_2$ (Prashanth et al. 2015), Al-based BMGs (Mu et al. 2009; Yang et al. 2009, 2010; Yan et al. 2011), Zr-based BMGs (Balla et al. 2009, 2010; Yang et al. 2012; Zhang et al. 2015; Harooni et al. 2016), and biomaterials and implants (Gibson et al. 2010b; Wang et al. 2016c) have been processed by SLS/SLM. Interested reader is referred to cited literature.

SECTION 3

MODELING AND SIMULATION

3.1 WHY MODELING AND SIMULATION?

Although in use since ancient Roman times (Rafique 2015), modeling and simulation picked up interest and achieved pinnacle in modern-day scientific and engineering sectors with the advent of computer technology which came not more than two decades ago. Now, it has proved itself to be an important integral part of product and part design, product development as well as prediction, utilization, and enhancement of properties. Various branches of modeling and simulation, ranging from part-scale modeling that involves development of codes of theorems in advanced computing platform such as Java, C, C++, and MatLab Simulink to their simulations in customized simulation packages such as Solidworks, Ansys, and Catia to performing complex atomistic simulations in dedicated software, have now become an integral part of design procedure in major industrial clusters. Its use in research and development is also becoming an important part of whole process to eliminate so-called trial and error methods that are waste of not only time but also energy, materials, and resources. In materials science and engineering mainly two of its branches are routinely used. These are "part-scale modeling and simulation" and "atomistic modeling and simulation." The former is used for the complete design of complex machinery segments, equipment, assemblies, subassemblies, their materials of fabrication, and property prediction in different regions as a function of extrinsic parameters such as heat, velocity, pressure, and time, while the latter is used for the prediction, estimation, and improvement in atomic-scale properties using theories of atomic configuration and arrangement mainly relying on intrinsic parameters (such as specific heat/latent heat, heat of fusion). The unique ability of atomistic modeling and simulation is that it uses atomic functions and their variables to generate knowledge about their behavior under various impulses. In both cases, the use of these methods is a big help and support in saving time and materials resources, as well as improving functionality and property enhancement.

3.2 CAPABILITIES/POWERS AND LIMITATIONS

The exponential rise in the use of modeling and simulation with the advent and progress of computer technology and increase of computing power of machines gave rise to much easiness in the design and development process. Many difficult or, in some cases, impossible to envisage problems can now be simulated using these computing platforms. These include simulation of water flow and its patterns in rivers and channels, simulation of the interior of sun, stars, and other heavenly bodies, cosmic events, and nuclear engineering problems. However, despite these advantages, there are still situations and applications that limit the use of modeling and simulation techniques. These include unavailability of strong efficient computing algorithms (with lesser approximations) needed for the replication of actual real-world situations, unavailability of real-world experimental data (physical constants) needed to simulate a particular problem, unavailability of more accurate deterministic or nonprobability-based models using actual situations rather than basing their outcome on probability. Owing to these reasons, there is still need for further investigation and removing bottlenecks from modeling and simulation problems.

3.3 TYPES OF MODELING AND SIMULATION

Modeling and simulation techniques can be divided into the following types depending on how the process is carried out and result attainment is sought out. The brief description is given below.

3.3.1 ENERGY MINIMIZATION

In essence, energy minimization also known as energy optimization or geometry optimization can be described as a set of numerical methods to find lower potential energy surface/state starting from state/surface of higher energy. These are extensively used in chemistry, physics, mathematics, metallurgical engineering, chemical engineering, mechanical engineering, and aerospace engineering to find stable/equilibrium states of molecules, solid, and items. Extensive studies have been carried out in various fields making use of energy minimization techniques to formulate models highlighting the importance, significance, and use of this method in modeling and simulation and solution of engineering problems. Its various submethods include Newton–Raphson method, steepest descent method, conjugate gradient methods, and simplex method.

3.3.2 MOLECULAR DYNAMICS SIMULATIONS

Molecular dynamics (MD) simulation is another powerful method of solving physical movements of atoms and molecules. It heavily relies on Newton's equations of motion for solving a system of interacting particles in which their motion is calculated by interaction in a set period of time. Various authors have applied these techniques for solving various types of phenomena including heat transfer (Maruyama 2000, 2003), heat transfer coupled with phase change during laser–matter interaction (Wang and Xu 2001), modeling and simulation of mechanical properties of polycrystalline materials (Krivtsov and Wiercigroch 2001), and problems of hard and soft matter physics.

3.3.3 MONTE CARLO SIMULATIONS

Monte Carlo (MC) simulations are a set of methods that rely on random sampling to compute their results. They are often used in computer simulations of physical and mathematical systems. These methods are typically effective in calculating solutions of systems with many coupled degrees of freedom such as fluids, disordered materials, strongly coupled solids, and cellular structures. Their method of operation is based on handling many systems at one time. However, they are not "many-body systems" which are famous ways of solving atomistic problems. They differ from the latter in a sense that atomistic methods deal with solving potentials and functions of atoms and their electronic states which can be modeled using MC techniques (Broecker and Trebst 2016; Hao et al. 2016) but are not an integral part of these calculations. Some of the prominent examples detailing MC methods are solutions of heat transfer problems (Planas Almazan 1997; Howell 1998; Zeeb 2002; Modest 2003; Frijns et al. 2004), microstructure prediction (Westhoff et al. 2015), carbon materials (Masrour and Jabar 2016), and nucleation phenomena (Binder and Virnau 2016). Recently, they have also been applied to studies of bulk metallic glasses (BMGs) (Guo et al. 2016; Nandi et al. 2016).

3.3.4 MISCELLANEOUS METHODS

There are further miscellaneous methods that are used for the modeling and simulation of various physical and engineering phenomena. Some of the prominent ones are Langevin dynamics (LD), normal mode (harmonic) analysis, and simulated annealing (SA). *LD* methods are used for the mathematical modeling of dynamics of molecular systems. In summary, they can be explained on the basis of Newton's second law of motion by adding two force terms (one for frictional force and other for random force) to approximate the effect of neglected phenomena. Their examples are solution of protein folding problem (Antonie 2007) and BMGs (Snook 2006). *Harmonic analysis* is used for simulation in which the characteristic vibrations of an energy-minimized system and the corresponding frequencies are determined assuming its energy function is harmonic in all degrees of freedom. Normal mode analysis is less expensive than MD simulation, but requires much more memory. Similarly, *SA* is a random-search technique that exploits an analogy between the way in which a metal cools and freezes into a minimum energy crystalline structure (the annealing process) and the search for a minimum in a more general system; it forms the basis of an optimization technique for combinatorial and other problems (Busetti 2003). Their use to BMG problems exists in a few cases but bears a prominent potential keeping in view the complexity of the problems encountered in BMG systems.

3.4 MODELING AND SIMULATION OF NUCLEATION (MICROSTRUCTURAL EVOLUTION) IN SOLIDIFICATION

Modeling and simulation of nucleation and growth (NG) (microstructural) evolution processes has been extensively studied since the early days of the development of theories of modeling and use of computer technology for simulation. Initially, linear analytical/deterministic approaches were adopted to model simple transport phenomena pertaining to heat and mass transfer in one dimension only. Development of Classical Nucleation Theory (*Christian,* 2002) and its application to BMGs (Gránásy 1993) is an excellent example about modeling of nucleation phenomena. They were quite useful in explaining NG in one dimension. However, they suffer from three main drawbacks: (a) Linear, one-dimensional (1D) models cannot explain actual nucleation phenomena happening in three-dimensional (3D) bulk of alloy liquid leading to evolution of solid in bulk. (b) These linear models are not good at explaining nonlinear nature of transient processes, especially heat transfer which is a strong function of changing process parameters itself (and the evolution of new front is dependent on new properties in preceding step). (c) The intrinsic nature of solidification process is probabilistic. Analytical modeling (even in its transient state) gives exact answers without taking into account the time-dependent probabilistic nature of process; thus, chances of error exist. All these reasons lead to the evolution of probabilistic modeling which was not only tailored according to the need and nature of process but also takes into account the errors which might have resulted from changes in system variables and were not solely based on permutations and/or combinations. These models were 1D in nature initially, which evolved into 2D and 3D progressively as their understanding increased. One of the first successful reporting of 3D probabilistic model of nucleation dates back to 1993 at EPFL, Lausanne by Rappaz and Gandin (1993) (Figure 3.1).

In detail, they presented two models simultaneously incorporating mix mode modeling (combining the powers of deterministic with probabilistic modeling) occurring at different stages of solidification throughout the cross section of alloy casting. One was aimed at explaining the phenomena in two dimensions (Rappaz and Gandin 1993) and other in three dimensions (Charbon and Rappaz 1993). In this model, they used previously developed 2D models primarily aimed at solid-state transformations and used their own approximations to model 3D liquid-to-solid transformation phenomena (explained in detail in later sections). At the same time, another effort was made at Ecole des Mines de Nancy, France by Ablitzer (1993) to model different transport processes occurring in different metallurgical unit operations. However, the limitation associated with this method was the use of linear deterministic approach, which was satisfactory only to a limited extent. Different probability-based methods

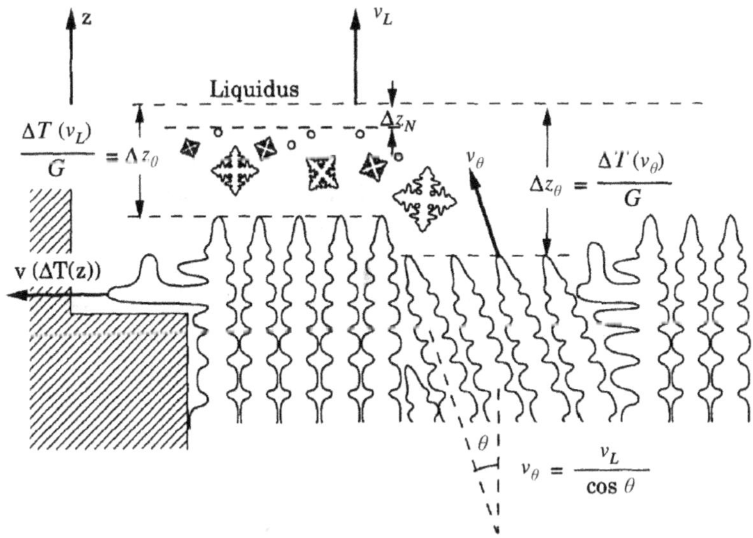

Figure 3.1. Schematic of various growth mechanisms occurring in a dendritic alloy with transient temperature conditions (Rappaz and Gandin 1993).

(phase field [PF], cellular automation [CA], and front tracking method) evolved simultaneously (aimed at solving different problems [predicting grain size, determining nucleant size, dendritic arm size, interdendritic arm spacing]) as understanding of solidification process gets deeper and deeper but in almost all approaches, MC-based simulation was applied to solve complex phenomena due to its rigor and power for solving multiphysics phenomena in liquids, solids, and gases (as described in Section 3.3). Out of these, CA-based approach remained successful to a large extent. These picked up pace and became a topic of interest in the scientific community when it was coupled with finite element approaches giving birth to so-called cellular automation finite element (CAFE) methods (Gandin et al. 1999; Guillemot et al. 2004; Chen 2014; Chen et al. 2014). Now, these are the most important topic of research due to its versatility and ability to model solidification processes to an extremely fine detail and at a scale which is very close to actual solidification patterns and microstructures observed in real alloys (Kremeyer 1998; Hebi and Sergio 2009; Zaeem et al. 2012; Zhang et al. 2012; Tian et al. 2016). This will be explained in detail in Section 3.4.

3.5 LARGE-/PART-SCALE MODELING

Large-/part-scale modeling involves modeling and simulation of phenomena happening at part level. It involves application of mathematical modeling to whole surface or volume of part. These usually include large and complex 3D algorithms as applied to explain evolution of certain phenomena (heat transfer, mass transfer, stress patterns, and fluid flow) happening at the scale of whole body. These may include application of more than one algorithm at one time in conjunction with the other (multiphysics phenomena). It usually involves the division of whole part in certain specific small segments and then applying the model (carrying algorithm(s)) to each individual segment. All segments are then combined to generate a synergic effect which gives rise to evolution of solution of certain problem as a whole. This involves computational and/or analytical (numerical) integration. In essence, its description and detail are described.

3.5.1 ANALYTICAL MODELING

Analytical modeling usually consists of finding a numerical analytic solution of complex mathematical equations using standard mathematical or statistical techniques. These techniques further imply the use of conventional or (nonconventional) modern mathematical methods. Models generated using these techniques are usually 1D or extension of 1D to 3D using simple approximations. Further, these could be linear or nonlinear. A computing platform is unable to read mathematical solution of, or even base, calculus equation. It is necessary to translate this equation to simple analytical solution using numerical methods. This gave rise to numerical analytical modeling. These include Newton–Raphson method, Runge–Kutta method, and other approximation methods. There are further two types of analytical modeling: (1) deterministic/continuum models (2) stochastic/probabilistic models.

Note: Use of "analytical" or "numerical" terms is dedicated to two different forms of solutions. The former is referred to finding a solution using standard mathematical equations such as those based on Newton's Laws of motion, Einstein's theory of relativity, Law of Gravitation, while the latter is referred to finding the solution of complex calculus or partial differential equations (PDEs) using more advanced or recent integration techniques which in essence translate an integer or PDE to analytic form which a computer can understand such as Newton–Raphson method, Runge–Kutta method. However, sometimes, these are synonymously used as well. For the sake of the present manuscript, these are referred to as "numerical analytical model" for "analytical model" and "numerical probabilistic model" for "probabilistic models."

3.5.1.1 Deterministic/Continuum Models

Deterministic/continuum models are models in which no randomness is involved in the development of future states of the system. Thus, a deterministic model will always produce the same type of results irrespective of initial state or start point. In other words, a deterministic model will always produce exact solution. Most of the partial differential equations (PDEs) describing physical phenomena in mathematics and physics are deterministic in nature. A good example is Schrödinger equation whose solution is exact in nature despite the relation between its wave function (ψ) and properties may have probabilistic nature. A typical example of nondeterministic (analytical) model/equation is Heisenberg's uncertainty principle.

3.5.1.2 Probabilistic/Stochastic Models

Stochastic/probabilistic models are models in which little or some randomness is always involved which is depicted in the final result. Thus, a set of values and initial parameters will always lead to a group of outputs, all or some of which are true in those conditions. The examiner has the liberty of choosing any of them as per their requirement (Krone 2004). Probabilistic models are not exact but are based on best approximation. Based on the hierarchical design of computing memory and power (binary 0101-type sequencing), it is easy for the computer to understand these models. In other words, it can be said that probabilistic models are based on statistical analysis of a problem. Also, their dependence is largely based on taking account of random change of states (i.e., in these models state function also keeps on changing). Although this can be explained by deterministic models as well it is much better to compute problems encountering randomly changing state variable in terms of probabilistic models. They are best to explain complex phenomena as they divide the problem into small group sets which eventually leads to group of results out of which best possible output could be generated. For example, application of wind force to tall structures is best described by probabilistic/ stochastic models as the amount, type, nature, and direction of wind load keeps on changing with time (time-based evolution studies). Similarly, earthquakes and design of earthquake-resistant structures is best explained by stochastic modeling.

3.5.2 COMPUTATIONAL MODELING

Computational modeling, as the name suggests, is a technique or a group of techniques in which rigor and strength of computing power is used to model and simulate a numerically translated analytic or probabilistic equation carrying an embedded model. Computational models and their use in predicting actual real-world behavior of complex systems is much more effective way of solving or arriving at near perfect solutions. Computational models combine the power of both deterministic and probabilistic models in few cases (mix models) to yield a much better result. The only impulse behind using these models is that they effectively employ and can incorporate even tough complex nodes which can only be solved using superior data handling and rationalizing power of computer rather than human brain. The latter might take months to solve an equation at all nodes of system while a strong computer only requires a few minutes to arrive at the same conclusion. In addition to that, the ability of a computing cluster to handle complex multiphysics randomly occurring simultaneous (parallel) phenomena at one time gives rise to their popularity in modern-day science and engineering. Some good examples for these models are finite element models, finite difference models (FDM), Phase field (PF) models, and Cellular Automata (CA) approach.

3.5.2.1 Lattice Boltzmann Method

Lattice Boltzmann method (LBM) involves solving basic "continuity" and "Navier–Stokes" equations and simulating complex fluid flow situations. LBM is based on microscopic models and mesoscopic kinetic equations. It originally has its roots from Ludwig Boltzmann's kinetic theory of gases. The basic idea behind these methods is that gas/fluid can be perceived as a huge volume consisting of randomly moving small particles. The exchange of momentum and energy between these is achieved through head-on, one-to-one collision (just as explained by Brownian motion). Thus, in some cases, these could be viewed as FDM for solving Boltzmann transport equations. In the best case scenario, if proper operators are chosen, LBM can even be used to recover Navier–Stokes equations (Bao and Meskas 2011). In metallurgy, alloy solidification, or microstructural evolution problems, it typically takes into account the fluid flow, distribution of liquid in microchannels, solute segregation based on its partitioning, dendrite evolution in 3D volume, its morphological development, interdendritic arm spacing, and distribution of liquid in that space. It typically considers a fluid volume element which is cohort of particles represented by a particle velocity distribution function for individual fluid component at each grid point. These particles are considered to be distributed under constraint conditions (i.e., under the action of applied force) and, as described earlier, are under free random movement (Brownian motion). Another condition governing their movement, existence, and evolution is that time average motion of particles is consistent with Navier–Stokes equations. Based on these parameters, very recently some studies have been carried out to model alloy solidification behavior in various types of alloy systems under different conditions (Raabe 2004; Sun et al. 2009, 2011a, 2011b; Asle Zaeem 2015; Eshraghi et al. 2015). Majority of these studies yielded precise and accurate results as they are conducted in parallel sequence (employing parallel computing techniques) combined with 3D coupling of other modeling methods such as CA (Eshraghi et al. 2015). For example, in a study conducted by Eshraghi et al. (2015) a new LB–CA model was created to simulate dendritic growth during solidification of binary eutectic alloys. The LB method was used to solve the transport equations and a CA algorithm was employed to capture the solid/liquid interface. The strength of the model was that it was able to capture evolution of thousands of dendrites in a macroscale domain with approximately 36 billion grid points in 1 mm^3 region. The parallel model showed a great scale-up performance on up to 40,000 computing cores and an excellent speed-up performance on up to 1,000 cores. This was truly a great achievement in terms of modeling complex phenomena in large-scale macroscale domains as it bears a potential to be adopted for and applied on to massively parallel supercomputers (e.g., Jaguar at Oak Ridge National Laboratories [ORNL], Deep blue at IBM and Cray at San Diago Supercomputing Centre).

3.5.2.2 PF Methods

PF methods are another most important class of computational methods which are used extensively for modeling and simulation of alloy solidification, microstructural evolution, and associated transport phenomena in various types of alloy systems under variant solidification and processing conditions. These have also emerged as an important simulation tool in recent years with the multifold advent of computers and increase in their ability to run various types of multiscale programs at one time (parallel programming). Below are some salient features of PF methods:

1. These methods revolve about finding a solution of PF parameter φ, which is a function of position and time; to describe whether the material is liquid or solid.
2. The behavior of this variable is governed by an equation that is coupled to equations of heat and mass (solute) transport.
3. The best way to explain "solidification pattern development" and "microstructural evolution" is to describe the evolution of liquid–solid interface in terms of smooth but highly localized changes of this PF variable φ between fixed values of 0 and 1 that represent solid and liquid.
4. One of the most powerful insights to the use of PF models is that it eliminates the use of boundary conditions. In normal solutions of problems, these boundary equations are required as they explain the evolution of liquid–solid interface in a particular confined volume in terms of solutions of transport equations (i.e., use of heat and mass [solute] transport by obeying thermodynamic and kinetic constraints). In PF calculations, these boundary conditions are not required. In these methods, the location of interface could be obtained by solving for numerical solution of PF variable at a position where $\varphi = \frac{1}{2}$.
5. Another big feature of PF methods is that they treat topology changes such as coalescence of two solid regions that come into close proximity (Boettinger et al. 2002).

In detail, PF methods can be classified into various intersecting classes: (a) those that involve a single scalar order parameter and (b) those that involve multiple order parameters. Another way is to divide them into (a) those derived from thermodynamic formulation and (b) those that are derived from geometrical arguments. Their elaborative discussion and explanation could be found in latest literature cited (Boettinger et al. 2002; Böttger et al. 2006; Emmerich 2009, 2011; Gránásy et al. 2014). They have also been applied to solidification and microstructural evolution problems in aluminum base (Zhang et al. 2006) and general (Nestler and Choudhury 2011) multicomponent alloys and glass-forming systems (Wang and Napolitano 2012). PF methods are rival to CA methods (Choudhury et al. 2012). Both can be used in conjunction with each other. However, their real strength lies in their use with complex finite element methods forming PF finite elements (PFFE) or CAFE, which will be described in a later section.

3.5.2.3 CA Method

CA is the most important, powerful, and latest development in computational models applied to solve the problems of transport phenomena happening in physical, chemical, metallurgical, and engineering systems. In some cases, this is called third-generation modeling approach. In essence, it involves dividing the bulk volume of system (material) into small numbers of finite cells, at the corners and boundaries of which the property of state variable and functions remains the same/constant. Some of the salient *features* of CA method are: (a) it is based upon strict physical mechanism (involving physical properties); (b) it makes use of low computation cost (thus benefits from low running cost), can be coupled with other models (explained below). A *limitation* could be that it may incur high initial (capital) cost (i.e., it may involve use of computing machinery with large processing and floating [RAM] memory).

First successful CA model was presented by Rappaz and Charbon at EPFL, Lausanne in 1993 (Charbon and Rappaz 1993). This model was built on previous 2D approaches of modeling solidification processes (primarily heterogeneous nucleation) and microstructural development and was extended in three dimensions by incorporating additional physical parameters. The power of the model also came from the fact that it takes into account the crystallographic orientations as well. Essentially, it comprises the following steps (Rappaz and Gandin 1993):

1. Grains and their properties (especially location and crystallographic orientation) are chosen randomly among a large number of cells and a certain number of orientation classes, respectively. For example, in case of pure cubic metals (e.g., B2 CuZr equiaxed dendritic

phase in glassy matrix in Zr-based BMGMCs), the parameters of interest taken into account were growth kinetics of dendrite tip and preferred (100) growth direction (which is already well-known direction of heat flow and thus growth in these crystals).

2. Further modeling then involves extending the application of initial model (which was applied to small-scale region of solid) to whole volume of casting (solidifying melt). In other words, the transport equations which were applied firstly to small finite solidifying volume are extended to whole bulk by applying them step by step to a large number of small regions covering bulk volume.

3. Constant temperature (usually ambient (T_r)) may be selected for steady state case.

4. Initial temperature (I_o) may be selected for transient case.

5. Grain growth morphology effects:
   a. Columnar-to-equiaxed transition (CET);
   b. Columnar zone morphology (which consists of selection and extension of columnar grains forming columnar zone); and
   c. Impingement of equiaxed grains (giving rise to their size reduction and fineness) is also taken care of.

6. Final step with respect to effect of the alloy concentration and cooling rate which affects final evolved microstructure is verified by rigorous experimentation.

The model has been specifically used by combining it with finite element models (CAFE) (Gandin and Rappaz 1994; Gandin et al. 1999; Guillemot et al. 2004; Chen 2014; Chen et al. 2014) as well as individually for the explanation of various metal and alloy systems (e.g., Mg alloy AZ91 [Hebi and Sergio 2009], brass [Tsai and Hwang 2011], Al-Cu Systems [Choudhury et al. 2012], ordinary binary alloys [Kremeyer 1998], Fe-based multicomponent alloys [Wang et al. 2016b], TC4 alloy [Tian et al. 2016], hexagonal crystals [Zaeem et al. 2012], and other multicomponent systems [Zhang et al. 2012]). The method has also been used for prediction of alloy microstructure in a specific process such as directional solidification (Wei et al. 2014), laser deposition (Tian et al. 2016), continuous casting (Bai et al. 2016), and additive manufacturing (Zhang et al. 2013a). A powerful advantage could be achieved by combining CA with PF methods, but this approach is still in its nascent stage (Zaeem et al. 2012).

3.5.2.4 Miscellaneous Methods

These include methods that employ techniques involving use of special algorithms and computing strategy. They have their own advantages but with respect to complexity, mass volume of castings handled, and type, shape, size, and morphology of melt pool encountered in laser-assisted and other additive manufacturing (electron beam) processes, these are rarely employed. Their special features and tailored abilities can be used in special circumstances and solving special problems. Some of the examples of these are virtual front tracking (VFT) model and sharp interface model. These are explained as follows.

3.5.2.4.1 VFT Model

This is a quantitative model which has specialty that it can be used to explain dendritic growth in systems which have low Péclet number (ratio of thermal heat convected to fluid [externally] to thermal heat conducted within fluid) systems. It is best when it describes thermal transport in two dimensions. In this case, it can even handle complex multiphysics fluid with high viscosities (liquid melt with pronounced mushy zone) (Zhu and Stefanescu 2007). Its variant (3D model) exists but computation becomes more complex and lengthy putting load on computing machinery and its power. A solution to this problem may be achieved by employing them only to less viscous (gas and liquid [petroleum-based]) systems (Chen et al. 1997; Dijkhuizen et al. 2010), but this is beyond the scope of solving for metallurgical problems (not the aim of present research). An excellent example of the use of this method is given by Zhu and Stefanescu (2007), in which they solved solutal transport equation to generate data obtained by local temperature, curvature, and local actual liquid composition to arrive at difference between the local equilibrium composition necessary to describe kinetics of dendritic growth. By employing this strategy, the dynamics of dendritic growth (including noise-free side branching) from initial unstable (fluctuating) stage to steady state stage was accurately predicted. The biggest advantage of adopting their technique was that it made the model completely meshless when going for simulation in a computing platform. Further efficiency in decreasing the computational time was achieved by calculating dendritic growth directly from fraction solid (rather than initial growth velocity calculations). This was a big achievement and a huge advantage of using the thermal transport (mass transfer) data directly to arrive at geometrical features of growing interface (dendrite). Another excellent example of coupling is given by Song et al. (2014) in which they combined CA model with VFT to describe crystallographic preferred orientation-based β- to α-phase transformation of TA15 alloy at constant temperature (isothermal). They used both techniques to achieve solid-state transformations in this multiphase alloy. This again is an excellent way forward to use various powers of different models to arrive at solutions of complex phenomena. However, more research is needed to optimize this regime.

3.5.2.4.2 Sharp Interface (SIF) Model

These models again, as the name suggest, are a method or a set of methods which incorporate interface phenomena and evolution of a certain ("liquid–solid" or "solid–solid" [Wang et al. 2015]) interface as a function of time. Owing to these reasons, these can best be used to explain evolution of interfaces during solidification as well as during heat treatment/solid–solid transformations. However, they have unique features of incorporating various auxiliary effects (such as surface energy, sharp moving interface, and velocity) while aiming at the solution of final equilibrium state and its growth. Vermolen (2006) and Vermolen et al. (2006) used this method to solve diffusion equations for dissolution of second-phase solid particles in multicomponent alloys taking account of surface energy effects. This was a unique work as this was the first attempt of its kind to explain dissolution in multicomponent alloys and was successfully explained by the use of SIF model. They also described various other modeling suites developed previously which base their simulation strategy on SIF models and their improvement, namely the model by Ågren (1982), Caginalp and Xie (1993), DICTRA (which simplified the complex problem of boundary conditions at the interface to a hyperbolic relationship). Recently, these models have also been applied to solve NG in electrodeposition (Nielsen and Bruus 2015) by using a combination of continuum and random noise (perpetuating SIF) terms in final equation. This again is an excellent example to combine the powers of two different modeling regimes. This trend is proving successful and is on the rise to solve difficult problems in complex situations (Sato and Ničeno 2013; Francois et al. 2006).

3.5.2.5 Mix Mode Modeling (Analytical and Computational Models)

These topics have already been introduced in previous sections (Guillemot et al. 2004; Francois et al. 2006; Sato and Ničeno 2013; Chen 2014; Chen et al. 2014; Song et al. 2014) so it will not be described in detail here. Merely, an introduction will be given to create a sense of their operation and applicability. In essence, mix mode modeling and simulation approach are based on combining the powers of two different models under one umbrella. This necessarily does not involve their computation in parallel with each other in a so-called *parallel programming* on computing clusters (also known as supercomputers). However, if the latter is done, it yields much better solutions of complex problems in much less computing time and using much less computing power than what would have been possible if the solution is run in standalone mode on single computing machine. Essentially these can be divided into two main types: (a) *Deterministic— PF models* and (b) *Deterministic—CA models*. Their detailed description depends on individual branch of engineering or physical science to which they are applied and, on the researcher (or research group) that is/ are responsible for their development. Popular regimes which now have become accustomed to be used toward solution of most engineering problems related to multiscale/multiphysics transport phenomena are CAFE modeling, PFFE modeling, and CA–VFT regimes.

3.6 ATOMIC-SCALE (ATOMISTIC) MODELING

Unlike part-scale modeling, another most powerful yet versatile method of predicting properties, phase change, their evolution and behavior is by the use of *atomic scale* or popularly known as *atomistic modeling*. Atomistic methods involve the use of special theorems developed for solving atomic-scale functions of all atoms in a given volume of material. Their functions are solved in terms of individual as well as interacting potential of each atom. This gives final results representing properties of materials in terms of solution of their functions. Atomistic modeling techniques use modern superior computing powers to explicitly arrive at the combined solution of every atom in system. This in detail can be done using various techniques specific to atoms or set of atoms (e.g., MD, MC, first principle [ab initio], force fields [interatomic potentials] with or without approximations). As interacting atoms are the foundation stone of all the materials, atomistic modeling has helped enable a new field of determining the properties of materials known as "computational materials studies." Some of the *advantages* of atomistic modeling and simulation are (a) they greatly help in reducing the cost of experimentation, (b) a huge saving in use of materials may be achieved, (c) process efficiency may be increased, (d) overall time in running experiments (down time) is reduced. These abilities and advantages help in (a) discovery and (b) development of deeper understanding of varied, difficult, and unexplained phenomena of materials. Despite this, one of the biggest *challenges* posed to atomistic modeling and simulation is inherent difficulty to handle multiscale, multiphysics phenomena which spans over many atoms (of changing wave functions) at one time. Another important area in which atomistic modeling has potential use is modeling with effective interactions between atoms, called interatomic potentials. These potentials do not treat the quantum nature of electrons explicitly, thus models employing them are extensively faster than quantum-based models. They might suffer from loss of some accuracy but still can model billions of atoms at once. Good examples of use of such techniques are nanocrystalline materials and friction between surfaces.

3.6.1 CLASSICAL MOLECULAR DYNAMICS (MD)

As the name suggests, MD calculations are done to understand the dynamics/motion of assemblies of molecules in terms of their structure and the microscopic interactions among them to understand their properties at a very small scale. A guess is developed at atomic level between interacting molecules and is presented to accumulate an effort to obtain exact predictions of bulk properties. History of MD techniques and their development dates back to 1957 when Alder and Wainwrigth first (1957) introduced the concept of interacting hard spheres and their phase transitions. This was based on original work to develop MC methods in 1953 (next section). A unique feature of MD techniques is that solutions generated are exact in nature such that they can be made as accurate as desirable. Only limitation is posed by computing power which is associated with capital and running budgets. Another feature of these calculations is that hidden dynamic details behind bulk measurements can be revealed simultaneously without lapse in time step. An example of MD is the link between the diffusion coefficient and the velocity autocorrelation function, with the latter being very hard to measure experimentally, but the former being very easy to measure. Ultimately, the aim is to arrive at direct comparison with experimental measurements made on specific materials, making it absolutely essential to have the explanation of phenomena in terms of good model of molecular interactions. The overall aim of MD is to reduce the guess work and fitting to arrive at ensemble of minimum. In addition, MD techniques can also be used to explain difference between good and bad or just to generate a comparison based on generic data in which case it is not absolutely necessary to have perfectly realistic molecular model but simple assumptions based on fundamental physics suffice. Very recently, MD simulations have been used for modeling of atomic-scale phenomena and BMGs spanning from understanding of structure (Amokrane et al. 2015; Tang and Wong 2015), explanation of indentation-based deformation in bulk (Wang et al. 2010c; Wu et al. 2015), transition from elasticity to plasticity (Feng et al. 2015), brittle fracture (Falk 1999), general mechanical (stress–strain behavior) properties (Krivtsov and Wiercigroch 2001), to evolution of structure in laser–matter interactions (Wang and Xu 2001).

3.6.2 MONTE CARLO (MC) SIMULATIONS

Monte Carlo techniques are another set of procedures used for explaining atomic-scale phenomena in terms of a set of probability-based possible outcomes out of which one could be chosen at random. History of MC techniques evolved with MD techniques and can be dated back to 1953 when Metropolis, Nicholas and coworkers (1953) at Los Alamos National Laboratories (LANL) developed a method for solving state equations (encompassing interacting molecules) by fast computing machines. It was later modified in 1957 by Wood and Jacobson (1957) again at LANL and its implications are still going on in modern-day science and engineering. The power of MC calculations is that it can explain various types of phenomena and processes. For example, it can be used to describe and account for *sampling* from a given bulk. The objective in this kind of approach is to gather information about a random object by observing many realizations of it. In this kind of scenario, random physical mimicking is done to observe the behavior of some real-life systems such as production line or network of conglomerated materials. Another example of the use of MC simulations is *estimation*. In this kind of situation, MC techniques can be used to account for certain numerical quantities related to simulation model. An example in the artificial context is the evaluation of multidimensional integral via MC techniques by writing integral as the expectation of random variables. Yet another most powerful insight to MC techniques is their use for *optimization*. This perhaps is the most powerful, most varied, and most frequently used feature of MC techniques. In many applications, these functions are deterministic and randomness (which is a feature of probabilistic techniques) is introduced artificially in order to more efficiently search the domain of objective function. In best case scenario, MC techniques are used to optimize *noisy* functions, where function itself is random—for example, the result of MC simulation. This is the most varied feature used in solving materials science and engineering problems apart from *sampling* (Kroese et al. 2014). For the first reported time, it was used to explain phenomena and structure development in binary Ni–Nb BMG in 1993 (Pusztai and Sváb 1993). Very recently, it has been extensively applied to the problem of BMG and their structural and mechanical evolution. For example, Hu et al. (2010) used this technique to describe probabilistic dual-phase magnetic behavior of Nd-based BMGs. Similarly, it is used to describe nanoscale structure development and structural relaxations (devitrification) in Zr-based BMGs (Hwang et al. 2012). A similar phenomenological study was carried out in Ni-Ti-Mo-based BMGs (Yang et al. 2016). In another study, MC techniques were used to study surface modification of Zr-based BMG by use of low-energy Ar or Ca ion implantation for biomedical application (Huang et al. 2015).

3.6.3 AB INITIO METHODS/FIRST PRINCIPLE CALCULATIONS

Ab initio methods or first principle calculations are set of methods derived out of quantum chemistry which are used to determine the state of a given system using its functions. Its features are based on quantum Molecular Orbital (MO) theory. It does not rely on empirical values of system but is based solely on established laws of nature (Gilbert 2007). Solution of Schrödinger equation forms the basis of MO theory. It yields energies and orbitals of electrons. *Energies* can be used to determine or expressed in terms of geometry optimization, reaction energetics, activation energies for kinetics, and UV/Vis absorption prediction, whereas *orbitals* can be used for graphical display (including assessment where reactant might attack), charges, dipole moments, electronic potentials, and NMR shieldings. In some cases, this is synonymous with MO calculations. History of these methods goes back to 1950 when Robert Parr first introduced their use to solve MO calculations of the lower excited electronic levels of benzene (Parr et al. 1950). Since then, the scientific world has seen many folds increase in their development. Most of these developments came as a result of (a) dramatic increase in computing speed as well as due to (b) design of efficient quantum chemical algorithms (Gilbert 2007).

3.6.3.1 Advantages

- They are easy to perform.
- Calculation cost is less (calculations are cheap in terms of their operation). However, experimental cost (to prove them) is high.
- Calculations can be performed on any system (even those that do not exist) while experiments are limited on only highly stable molecules (mostly monovalent or divalent).
- Calculations are safe (as merely it is mathematical procedure involving paper and pen or at most a computer) while setting up an experiment is very risky and dangerous task (installation, operation, control and maintenance of large number of parts, equipments, and machinery).

3.6.3.2 Disadvantages

- Calculations are too easy to perform by use of software (rendering them less use of human knowledge).
- Time to carry out calculations can be very lengthy (especially if used on low-power computing machinery).
- Calculations can even be performed on hypothetical (imaginary) systems (which are waste of time).

3.6.3.3 Procedure to Carry Out ab initio Calculations

In essence, an ab initio calculation involves selection of (a) a method and (b) a basis set. Some of the examples of *methods* include

    a. Hartree–Fock (HF) method and Slater determinant
    b. Car–Parrinello method and
    c. Density functional theory (DFT)

For example, *HF method* is the simplest wave function-based method. It forms the foundation for more elaborative electronic structure method (e.g., self-constrained field [SCF], DFT). It posits a trial form of the N-electron wave function and uses the variational theorem (which is wave function-based approach using mean field approximation [Gilbert 2007]) to obtain an approximate solution. The basic idea here is to express the wave function as a product of individual spin orbital solutions, the so-called *Hartree product*. This form of trial wave function does not obey the indistinguishability requirement for fermions, that the swapping of two particles generates an identical but negated wave function: since in general all of χ functions can be different. In order to solve this enigma, instead of Hartree product, a similar multiplicative combination of individual spin orbital that forms the negative upon swapping is used. For an arbitrary number of electrons, this can be achieved using determinant of a matrix, called *Slater determinant*. Similarly, *basis set* (which is second part of performing ab initio calculations) is a set of atomic orbitals (AO), whose coefficients are to be determined using *HF* method to make Molecular Orbitals (MOs).

3.6.3.4 Approximations

Schrödinger equation can be solved exactly for H atom only to find solutions of MO theory (in terms of AO). So, all other AO solutions are approximations. Some of these approximations are:

- *Born–Oppenheimer* approximation
- *Independent electron* approximations
- *Linear combination of AO* approximations

3.6.3.5 Evolution of HF Theory

With the passage of time, HF method evolved into more simplified methods to find solutions of MO theory. In essence, they all involve solutions of time-independent Schrödinger wave equation which may be described as

$$\hat{H}\psi = E\psi \tag{1}$$

where ψ = wave function. It is postulate of quantum mechanics. It is a function of the positions of all the fundamental particles (electrons and nuclei) in the system. $\hat{H}$ = Hamiltonian operator. It is the operator associated with the observable energy. It contains all the terms that contribute to energy of system

$$\hat{H} = \hat{T} + \hat{V} \tag{2}$$

where $\hat{T}$ is kinetic energy operator and $\hat{V}$ is potential energy operator (Gilbert 2007).
E = total energy of system. It is the operator associated with the observable energy.

These evolutions are briefly described as follows:

- SCF method which evolved as a result of conjunction with HF method. Thus, it is sometimes called HF–SCF method.
- Møller–Plesset (MP) perturbation (MP1) theory which was next evolution of HF theory. In this, Hamiltonian is divided into two parts: $\hat{H} = \hat{H}_0 + \lambda\hat{V}$. Second term is perturbation and is assumed to be small. The energy and wave function are expanded as power series in λ (which is later set to unity). (Note: Even in solving MP1 theory, wave function and energy are HF wave function and energy.)
- MP2: It is next evolution of MP1 theory in which wave functions remain the same. However, energy is explained in terms of MP2 energy and bears accumulative treatment of wave function by use of summations.
- Since 1996, even MP2 method evolved into more simplified theory known as Density Functional Theory (DFT). The strength of DFT method is that this energy of system is obtained from electron density rather than from more complicated wave function. It is an approximation but very good to explain the energies of system.

Recently, this method has been extensively applied to solve various types of problems in chemistry, physics, materials science, and engineering. They have also been used to solve problems of BMGs and their composites (BMGMCs), in particular to study their structural evaluation (liquid state, its arrangement, and final configuration into glass) (Amokrane et al. 2015) and origin of glass-forming ability both in binary and ternary BMGs (Sha and Pei 2015).

3.6.4 INTERATOMIC POTENTIAL/FORCE FIELDS

Despite their rigor and power, MD and MC methods cannot be used directly on atomic systems to explain their interactions. Their application requires the adoption of some rules that govern the interaction of atoms in a system. It is convenient in classical as well as semiclassical simulations to express these rules in terms of potentials. These potentials are known as interatomic functions. Usually, they are represented by a potential function U or

$$U = U(r_1 + r_2 + r_3 + r_4 + \cdots + r_N) \tag{3}$$

where U is the sum of potential energies of system of N atoms. It is dependent on individual coordinates of atoms represented by $r_1, r_2, r_3, \ldots, r_N$. As explained in the previous section, some approximations are helpful in defining these potentials. For example, it is assumed that electrons adjust to new atomic potentials much faster than the motion of atomic nuclei (Born–Oppenheimer approximation). Similarly, forces in MD simulations are defined by potential $\vec{F}_i$ which is explained as follows:

$$\dot{\vec{F}}_i = -\dot{\vec{V}}_{r_i} U(\vec{r}_1 + \vec{r}_2, \ldots, \vec{r}_N) \tag{4}$$

In general, selection of a particular type of potential depends on following characteristics: *accuracy* (a measure of how well systems can reproduce properties of interest closest to required/desired values), *transferability* (accounts for ability of a system to allow the study of properties and generate results for which it is even deemed unfit), and lastly *computational speed* (defines how fast calculations can be done on a system to solve its potentials). Finally, the choice of interatomic potential depends on area of intended application. There are almost no "good" or "bad" potentials. Their choice and selection depend on type of problem in question. On the basis of this, they can be "appropriate" or "inappropriate." High accuracy is often the prerequisite for problems pertaining to computational chemistry while computational speed is often the bottleneck in materials science, in which case, processes have a collective character and require large time for their effective solution.

3.6.4.1 Classification of Interatomic Potentials

Interatomic potentials can be classified into two types on the basis of how they will be used for a particular problem. These are

- Pair (Two-Body) Potentials and
- Multi (Many-Body) Potentials

Explanation: Total energy of a system of N atoms with interaction described by an empirical potential can be explained in a many-body expansion.

$$U\left(\vec{r}_1, \vec{r}_2, ..., \vec{r}_N\right) = \sum_i U_1\left(r_i\right) + \sum_i \sum_{j>i} U_2(\vec{r}_i, \vec{r}_j) + \sum_i \sum_{j>i} \sum_{k>j} U_3(\vec{r}_i, \vec{r}_j, \vec{r}_k) + \cdots$$

(5)

where

U_1—one-body term, due to an external field or boundary condition (wall of container)

U_2—two-body term, or pair potential. (This interaction of pair of atoms depends only on their spacing and is not affected by the presence of other atoms.)

U_3—three-body terms arise when the interaction of a pair of atoms is modified by the presence of a third atom.

This leads to classification of system into two types: (a) pair potential (when only U_2 is present or calculation of potential is restricted to second term only) (Figure 3.2) and (b) many-body potentials (when calculation of potential involves use of U_3 or higher terms). This classification also makes it easier for the procedure by which potentials are chosen. It is briefly described below.

1. *A functional form for a potential function could be assumed based on which certain parameters are chosen to reproduce a set of experimental data.* This gives rise to so-called *empirical* potential function. Most of the time these are two-body or pair potentials (e.g., Lennard–Jones potentials, Morse potentials, Born–Mayer potentials).
2. *Electronic wave function for fixed atomic positions could be chosen.* This inherently is difficult for a system of many atoms. Different approximations (as described earlier) are used and analytical *semi-empirical* potentials are derived from quantum mechanical arguments. These typically are multi (many) body potentials (e.g., Embedded Atom Method proposed by Baska, Daw at Sandia National Laboratories in 1984 [Daw and Baskes 1984] and explained for fcc metals and alloys in collaboration with their colleague Dr. Foiles in 1986 [Foiles et al. 1986] and cubic metals in 1992 [Baskes 1992], glue method [Ercolessi et al. 1988], effective medium theory, reactive potentials, bond order potentials [TB potentials] [Donald et al. 2002]).

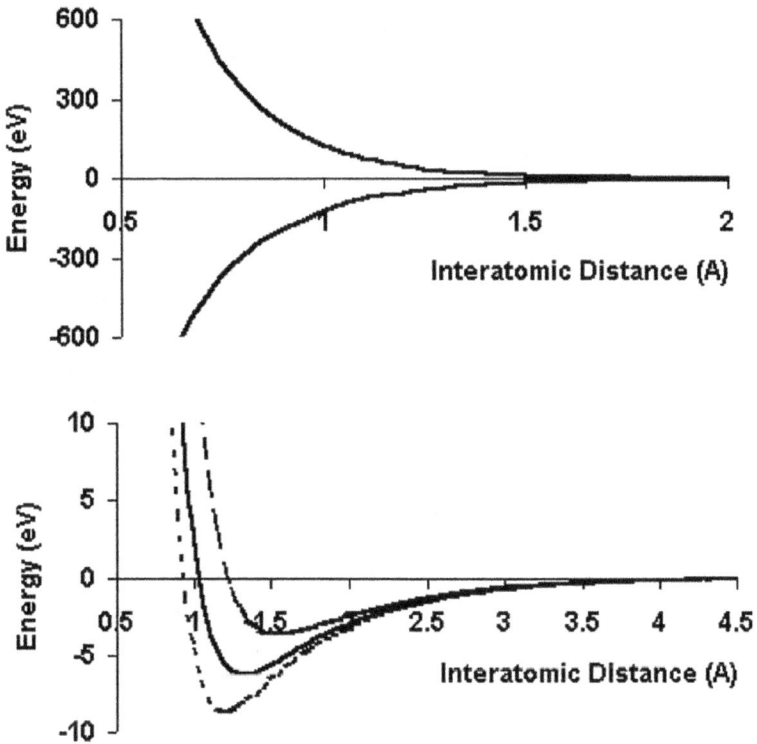

Figure 3.2. Plots of pair potentials: Top: Attractive and repulsive pair terms as a function of interatomic distance. Bottom: pair terms for triple bonds (dotted curve), double bonds (solid curve), and single bond (dashed curve) obtained by multiplying the attractive pair term by appropriate bond order value and adding it to repulsive pair term (Donald et al. 2002).

3. *Directly performed electronic structure (quantum mechanics based) calculations of forces during so-called ab initio MD simulations (as explained earlier) could be chosen* (e.g., Car–Parrinello method using plane wave pseudopotentials).

Usually, *pair potentials* are performed for inert gases, intermolecular van der Waals interaction in organic materials, and investigation of general classes of material nonspecific effects, whereas *many (multi) body potentials* are performed for metallic systems. In addition, all other types of potentials have their own specific applications and are suited for the solutions of particular system (e.g., force field method can be used for covalently bounded systems, reactive potentials are for carbon and hydrocarbons as described by Brenner in his original derivation [Donald et al. 2002]).

3.7 VERY RECENT TRENDS/FUTURE OUTLOOK

An interesting (not yet adopted) approach would be to combine the power and rigor of both part-scale and atomic-scale modeling and simulation methods. This might involve the use of very powerful computing machines and strong parallel programming cluster but it is bound to yield unique and fascinating properties of materials as they are used in actual application which otherwise are not possible as atomistic models are limited to be applied to confined volume of materials (usually hundred or few hundred of atoms) while part-scale simulations are not concerned with phenomena happening at atomic or molecular level. They only yield data and predict macroscale behavior of part (and hence its material) as a function of external impulse (temperature, pressure, stress). In this regard, few efforts have been made to model the behavior of biological molecules (proteins) (Pronk et al. 2013), combination of transistor and its tunneling effect to form world's smallest transistor at ORNL Jaguar supercomputer (Seabaugh 2013), multimillion atom biological MD simulation (Schulz et al. 2009), etc., but still research in this field is very much in its nascent stage and restricted to large-scale well-developed computing clusters. More effort is needed to bring this technology to pubic and for its commercialization.

SECTION 4

Modeling and Simulation of Solidification Phenomena during Processing of BMGMC by Additive Manufacturing

4.1 INTRODUCTION

This section deals with the evolution of microstructure during processing of bulk metallic glass matrix composite (BMGMC) in incipient transient liquid melt pool formed in additive manufacturing (AM). Analysis is divided into two sections. The *first section* deals with evolution of melt pool as a result of interaction of highly localized, focused laser light with matter (metal powder). This results in the formation of melt pool whose shape, size, geometry, and transient behavior are very much a function of heat transfer coefficients (HTCs) evolving at every step of its formation (melting and homogenization) and dissipation (solidification). Solidification in this section is considered by modified general (classical) nucleation theory (CNT). Once formed, this pool travels as laser traverses its path all along the powder bed dictated by CAD geometry at the back-end. The *second section* deals with microstructural evolution during solidification, which is primarily *solute diffusion-* and *capillary action*-dominated phenomena. This is dealt with by microscopic 2D and 3D probabilistic cellular automation (CA) models which model nucleation and equiaxed dendritic growth resulting in the formation of microstructure in liquid melt pool as it solidifies (Note: only "vitrification [glass formation]" effects are taken into account). The evolution of microstructure is checked against variation of number density, size, and distribution of ductile phase in glassy matrix.

Inoculants for ductile phase formation were selected previously by edge-to-edge matching (E2EM) (Kelly and Zhang 1999, 2006).

4.2 MODELING AND SIMULATION OF HEAT TRANSFER IN LIQUID MELT POOL—SOLIDIFICATION

As microstructure formed during SLM is mostly columnar (Zheng et al. 2008), it is a good indicator that heat flux transfer from melt is highly unidirectional; thus, heat transfer from the bottom is a transient 1D process. Although heat is lost from material in x–y plane, that is, perpendicular to z-direction (perpendicular to build direction), its contribution is so low that it can be safely ignored. However, this was an old concept. New experimental observations have proposed a new concept according to which during SLM, a melt pool is formed. The "shape" of this pool is a function of

   a. Laser power (laser beam intensity)
   b. Presence of thermocapillary convection (Marangoni convection)

In even more advanced and recent models (King et al. 2015; Khairallah et al. 2016), transfer of heat after its generation is considered by three main parameters:

   a. Heat transfer due to convection
   b. Evaporation (i.e., formation of plasma) (this results in re-radiation [inverse radiation]) and
   c. Conduction from bottom and side walls

This is a very recent and advanced approach which, however, ignores Marangoni convection effects. Overall, heat transfer phenomena associated with solidification of metal in liquid melt pool in AM is associated with three processes:

   • Generation of heat (Laser–matter interaction)
   • Assimilation of heat (Melting and stages of solidification)
   • Extraction of heat

4.2.1 GENERATION OF HEAT (LASER–MATTER INTERACTION)

This is the first stage of AM in which heat is generated. The problem in this stage is related with impingement of light of certain intensity (I) on a solid surface for a certain amount of time which may result in production of heat. This interaction can be explained in terms of law known as "Beer–Lambert law."

4.2.1.1 Beer–Lambert Law for AM

Consider a thin layer of powder with thickness d_1, on a flat disk substrate of refractory metal with thickness d_2 and radius r uniformly illuminated by light of intensity I.

For absorptivity of powder (or melt) assuming uniform temperature throughout the disk, the temperature evolution is

$$(\rho_1 c_1 d_1 + \rho_2 c_2 d_2)\frac{dT}{dt} = A(T)I - Q(T) \tag{1}$$

where

$A(T) =$ Absorptivity
$Q(T) =$ Thermal loss (convective and radiative)
$I =$ Intensity
$\rho_1 =$ Density of powder
$\rho_2 =$ Density of substrate
$c_1 =$ Specific heat of powder
$c_2 =$ Specific heat of substrate
$d_1 =$ Thickness of powder
$d_2 =$ Thickness of substrate

Heat generated by this process is used for melt pool generation (its morphology, homogenization, and holding [generation of supercooled liquid (SCL) region and its progression]).

4.2.2 ASSIMILATION OF HEAT (MELTING AND STAGE OF SOLIDIFICATION)

As the heat generated above interacts with metal powder, it causes its melting and generation of liquid melt pool. Behavior of certain metal/alloy in melt pool can be explained by its cooling curve which is briefly described below.

4.2.2.1 General Form of Cooling Curve

Cooling curve of a metal/alloy is a plot of variation of temperature with time. It has different regions which embodied various types of information.

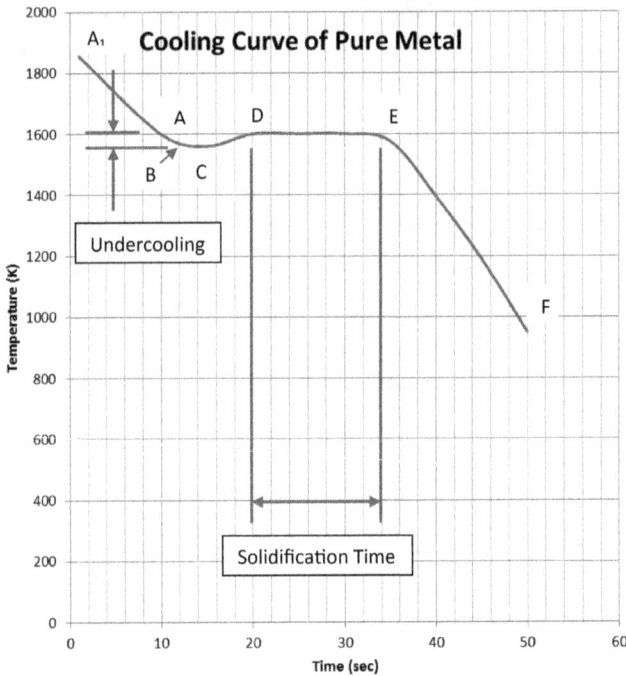

Figure 4.1. Cooling curve for a single-component pure metal (without any inoculants).

Cooling curve can have different shapes depending on metal or alloy type. A schematic cooling curve is shown in Figure 4.1 for a single-component pure metal (without any inoculants).

Its distinct regions are explained as follows.

Region above A_1: This is the region in which metal is in its complete liquid state and can be described by only melting and liquid state homogenization. Heat carried by metal in this region is "super heat" only and lost in the form of specific heat ($mc_p\Delta T$). This homogenization in turn depends on type of melting (gas/solid [coal]/liquid [oil] fired crucible furnace melting, electric [resistance/induction/arc] melting) and subsequent melt treatment. (Note: Homogenization is required by some external means in case of all modes of melting. Only induction furnace is manifested by self-homogenization due to phenomena of induction currents.)

Region A_1–A: This region is characterized by loss of superheat until first arrest point A. (Point at which first nucleant forms—explained in detail in later sections.) This is also called start of solidification. In pure metals it is sharp point (melting point) while in alloys, it can be a range (melting range). In BMGMCs/multicomponent alloys, it is also called start of SCL region. This region is followed by undercooling (ΔT_n) region which is described below.

Region A–D: This is the most important region of cooling curve (present case) for pure metals. In this region, metal cools down to a specific temperature characterized by a certain minimum amount of energy (activation energy for nucleation) needed to overcome a barrier of energy (energy barrier to nucleation) to create a liquid–solid (L–S) interface eventually leading to formation of stable nuclei out of melt. This region is further divided into two regions: A–C and C–D.

Region A–C: This is the region in which undercooling occurs, heat is extracted, temperature drops, and shape of cooling curve goes down. This is characterized by two energies described in above paragraph.

Region C–D: This is the region in which heat energy is absorbed, temperature is gained, and shape of curve goes up. This is called recalescence.

Notes:

a. Recalescence is gain in temperature as a result of thermal fluctuations caused by phase transformations occurring within solidifying melt/alloy.
b. Region A–C is characterized by another point: point "B" occurring in the middle of cooling curve. This is specifically shown in Figure 4.1 as intermediate point of SCL region. For present case model, transient heat transfer conditions will be modeled at this point as well to get better understanding of phenomena occurring in SCL in BMGMCs.

Region D–E: This is the region at which (after arrest point D), metal loses all its heat of fusion (mH_f). In this region transformation occurs at constant temperature in such a way that all liquid gets transformed into complete solid (all fine equiaxed grain formation at mold wall [Cu mold casting]/at surface of inoculant [heterogeneous nucleation—not present case], "equiaxed–columnar" transition, growth of columnar dendrites, columnar to equiaxed transition [CET] and growth of all equiaxed dendrites accomplishes). This is also called solidification time.

Region E–F: This is the region in which solid cools. That is, alloy reaches its completion of solidification and cools to room temperature in solid state. This again occurs after a sharp invariant point (point F) in case of pure metals and after a range in case of multicomponent alloys.

4.2.2.2 Cooling Curve for Well-Inoculated Zr-Based in situ Dendrite BMGMCs

Shape of cooling curve changes its form as melt is changed from single component to binary to multicomponent alloys. This can be explained in the form of various cases.

Case 1: Well inoculated single-component melt: In these types of alloys, undercooling/undercooled region (ΔT_n) diminishes and is almost absent. Inoculation with potent nuclei serves as active nucleation sites and triggers heterogeneous nucleation as the alloy reaches its first invariant point. Thus, no undercooling happens and solid alloy directly starts cooling as all liquid gets transformed to solid at constant temperature.

Case II: Binary alloys without inoculants (slowly cooled)
In these types of alloys cooling occurs in following steps:

1. Distinct undercooling occurs (characterized by drop and gain [recalescence] of temperature)
2. It is followed by region of constant temperature cooling which is called *local solidification*. This is only visible in case of very fluid alloys in which mushy region is very fluid/less viscous (not BMGMCs). This region is absent in most multicomponent (industrial) alloys as their solidification is dominated by mushy zone. (Note: BMGMCs are special case of alloys in which mushy region is extensively dominated but another phenomenon known as "sluggishness" governs the solidification. In these alloys, three laws [Hofmann and Johnson 2010] which describe BMGMC formation and evolution make sure that sluggishness not only dominates kinetics but also ensures "glass formation," that is, retaining SCL at room temperature).
3. Alloy solidification range (it depends on alloy. In slowly cooled binary alloys [most laboratory conditions], this is very clearly marked [usually bears an intermediate shape]).
4. At the end of this range, alloy becomes stable momentarily at constant temperature (usually negligible in most industrial castings) at which nuclei (dendrite arm branches) grow and fill interdendritic arm spacing and other small liquid pockets. This is marked by end of solidification. (In some cases, it is also characterized by start of CET and then growth of equiaxed grains.)
5. Following this point, solid alloy cools to room temperature or below room temperature (in case of cryogenic cooling).

Note: For theoretical analysis, cooling curve can be of any type of combination between type of alloy (single component, binary, and multicomponent), method of cooling (slow or fast), and inoculation (zero inoculation and well inoculated). All these can be drawn following rules of thermal transitions and kinetics. For simplicity and sufficiency, we will jump to cooling curve of multicomponent alloy (BMGMCs) fast cooled and well inoculated (present case).

Case III: Multicomponent alloys with inoculants (fast cooled) (present case BMGMCs)
In these types of alloys, cooling can occur following below steps:

1. No undercooling occurs (as there is sufficient amount [number] of potent nuclei which serve as sites for active nucleation triggering heterogeneous nucleation prior to loss of temperature [drop of cooling curve], and gain of temperature [recalescence—rise of cooling curve]).
2. This is followed by region of constant temperature at which all liquid gets transformed into solid. However, in these alloys, this region is very small (because of presence of marked mushy zone).
3. Instantly after this region, alloy enters in alloy "solidification range." As the alloy is very fast cooled, this region is again not very clearly identified which is typical behavior in case of fast cooled castings.
4. Following this, again alloy momentarily enters in brief constant temperature zone which marks starts of CET and growth of equiaxed grains (B2 CuZr phase equiaxed dendrites) until all liquid gets transformed into solid (end of solidification). This again is not very distinct as other phenomena (suppressing kinetics) dominate.
5. Finally, after this, BMGMC solidifies to room temperature.

Note: Shape of cooling curve in case of slowly cooled and fast cooled alloys is the slope of curve toward the end of cooling which is very steep in case of very fast cooled alloys (liquid melt pools [present case]).

4.2.2.3 Extraction of Heat—Determination of HTCs

In the development of model, HTCs will be determined at every point of cooling curve following earlier defined one-dimensional (1D) schemes (Rafique and Iqbal 2009). These will ensure time of solidification calculation during cooling following above cooling curve and helps in determining shape of melt pool and its transient behavior during cooling.

4.2.2.4 Final Time of Solidification

Final time of solidification is sum of time in each region/section of cooling curve of an alloy/melt. It will be determined using standard transport equations and will be used empirically to assess the conformability of AM process. Time of solidification gives other parameters as well such as fraction of mass solidified after a time *t* which is a direct measure of microstructure evolved during that time. It can be qualitatively (extrapolation) used to predict further (type [equiaxed, columnar, mix, and CET] and amount) evolution of microstructure with time.

4.3 MODELING AND SIMULATION OF NUCLEATION (HETEROGENEOUS) IN LIQUID MELT POOL— MICROSTRUCTURAL DEVELOPMENT

Modeling and simulation of microstructural development in liquid melt pool can be described by macroscopic and microscopic models of heat and mass transfer depending on type of alloy, its nature, number of elements, cooling curve, undercoolings (constitutional [solute/particulate], curvature, and interfacial), thermal and kinetic limitations, behavior of mushy zone, presence or absence of inoculants. These can be broadly divided into macroscopic and microscopic models (Rappaz and Gandin 1993) which are explained as follows.

4.3.1 MACROSCOPIC MODELS

By following the regimes of macroscopic models, finite element method (FEM) and finite difference method (FDM) can be used to explain microstructural development during both steady and transient state transport processes.

4.3.1.1 Limitations

Both FEM- and FDM-based models cannot fully describe mushy region, its behavior and evolution during solidification as they do not account for microscopic

- solute diffusion and
- capillary effects

which are primarily responsible for scale at which microstructure forms (which is very small as compared to macroscopic methods based on average continuity equations [Bennon and Incropera 1987; Voller et al. 1989; Ganesan and Poirier 1990; Ni and Beckermann 1991] in which it is assumed that solidification starts at liquidus and finishes at solidus/eutectic temperatures—a case of BMGMCs having good match of glass-forming ability and eutectic temperature [Ma et al. 2003; Lee et al. 2012]). In order to overcome these limitations, microscopic models were proposed.

4.3.2 MICROSCOPIC MODELS OF MICROSTRUCTURE EVOLUTION/FORMATION DURING SOLIDIFICATION

Stage 1 Model: These models take into account the mechanism of (1) grain nucleation and (2) grain growth in alloys which are solidifying with equiaxed dendrite or eutectic microstructures (Rappaz 1989). These do not account for alloys which are solidifying with columnar dendritic and planar interfaces. A modification of these accounts for equiaxed-columnar (at mold wall) and CET in bulk of liquid. (This will be discussed later.) These can be used to "describe microstructures" and "prediction of grain size" in case of eutectic compositions of BMGMC. Majority of these are based on "analytical/deterministic approaches" which can be described as follows.

4.3.2.1 Nucleation

- Choose a time "t" (initially nonzero value)
- At this time t, density of grains (which have nucleated in bulk) is a function of undercooling

$$d = f(\Delta T_n) \qquad (2)$$

$f(\Delta T_n)$ is difficult to be found from theoretical considerations alone. It needs to be found experimentally, that is, from a set of experiments.

Method 1 Measurement of cooling curve
This has been explained in detail in Sections 4.2.2.1 and 4.2.2.2.
Method 2 Measurement of grain density (optical micrograph of cross section [using ImageJ/manually]) for specimens solidified at various cooling rates (Rappaz and Gandin 1993)

4.3.2.2 Growth

As soon as grain has nucleated, and its growth can be explained by special modified case of CNT for BMGMC (a detailed treatment of modified CNT for BMGMC is given in Appendix A) and its distribution can be explained by constitutional supercooling zone/interdependence theory (propagation of L–S interface/L–S spherical front) (a possibility which is still under investigation by author for suitability for AM processes), it grows with an interface velocity which is also a function of undercooling.

4.3.2.3 Velocity of Growth

Velocity of growth may be written as

$$V_g = f(\Delta T_n) \tag{3}$$

In this case, there is no need to determine solidification kinetics of dendrite tip/eutectic (spherical front) interface by cooling curve or grain size but it can be determined by theoretical models developed (by using basic laws of physics) (Kurz et al. 1986; Trivedi et al. 1987) as applied to BMGMC only under transient condition.

4.3.2.4 Impingement

Impingement of grains as they grow is another important phenomenon which for all practical reasons governs the shape of grain after CET (CET in AM is recently explained by Basak et al. [2016] which is combined with present model and is explained in detail in Appendix B) and is mainly responsible for equiaxed dendritic grain formation, especially in eutectic composition which is assumed to be the case for present research.

This has been typically treated by

a. Standard JMAK (Avrami 1939; Christian 2002) correction or by
b. Geometrical (Price 1987; Zou 1989) or
c. Random grain arrangement models (Rappaz and Gandin 1993).

These "microscopic" solidification models have been coupled with "macroscopic" transient 1D heat flow calculations to successfully predict "microstructural features" specially "grain size" at the scale of whole process (part scale) (Thévoz et al. 1989; Stefanescu 2015).

4.3.2.5 *Limitations*

These deterministic models have their following limitations

a. **Grain selection**

   They cannot account for the "grain selection" which occurs

   a. Close to mold region/surface giving rise to columnar dendritic microstructure (in case of conventional Cu mold casting/TRC) or

   b. At surface of external inoculant particles (precursors of hetero-geneous nucleation) in case of well-inoculated melts (present case) giving rise to onset of columnar dendritic microstructure (at a very small length scale) since they neglect any aspect which is related to crystallographic effects.

b. **"Equiaxed–Columnar" Transition**

   They cannot predict the so-called equiaxed–columnar transition which occurs very near to mold wall (Kurz and Fisher 1986) or variation of transverse size of columnar grains (Chalmers 1970) (also known as columnar dendritic arm branching). This is explained in detail in individual cases for each type of metal (crystal structure)

   i. **Case 1: Cubic Metals**

   It is a well-established fact that for cubic metals, this "grain selection" is based upon a criterion of best alignment" of the <100> crystallographic axes of grain with heat flow direction (Kurz and Fisher 1986; Rappaz and Blank 1986; Chalmers 1970). Thus, this method cannot account for this anisotropic behavior of heat flow. A solution to this problem could be proposed by determining best fit direction by use of recent developments in crystallography and their application to solidification. *E2EM:* One way is to use E2EM technique at inoculant—ductile phase level (in case of Zr-based BMGMC) (present research). This gives rise to selection of suitable potent nuclei of certain size and specific preferred orientation (i.e., along a defined easy crystallographic plane (001)). If this crystallographic plane direction could be used in conjunction with macroscopic heat flow models, it can give rise to "prediction or selection of grain." In other words, if matching crystallographic axes (suitable for a potent inoculant selection for B2 ductile phase's preferred precipitation [in case of BMGMC]) could be best aligned with heat flow direction (or heat flow direction could be assigned to this preferred matching crystallographic axes) a best "grain selection" could be determined (one of the aims of the present research—not done previously elsewhere). This type of phenomena is particularly important in

    a. Directional solidification or

    b. Production of single crystal dendritic alloys for aerospace applications or

    c. Production of BMGMC by Bridgman solidification

Note: This is in addition to use of E2EM for selection of potent nuclei

ii. Case 2: BCC Metals

These methods are also ineffective in predicting "equiaxed–columnar" and then "branching of dendrite arms" in bcc metals (i.e., grain selection) as best alignment between heat flow and crystallographic direction is not known. Only assumptions are possible (i.e., in case of bcc best heat flow direction could be assigned to close packed direction).

iii. Case 3: FCC Metals

These methods are again ineffective in predicting the "equiaxed–columnar," "CET," and then branching of dendrite arms in fcc metals (i.e., grain selection) as best alignment between heat flow and close packed direction (111) could only be assumed (to a satisfactory qualitative level). More quantitative experimentation is needed to determine best directions along which heat flow occurred or revert to more advanced models.

c. Extension of a Grain into an Open Region of Liquid

They cannot account for extension of a grain into an open region of liquid.

d. Columnar to Equiaxed Transition

Finally, when very fine equiaxed grains at a region very close to mold wall/right at the interface of inoculant and melt are converted to columnar grains, which when grows, there comes a point/plane at which columnar grains get converted to not so fine equiaxed grains. This point is known as columnar to equiaxed transition (CET). These equiaxed grains finally extend toward the center of casting (wedge shape/melt pool centerline in case of AM). CET primarily happens as a result of thermal fluctuations which happen at melt (liquid) and solid (solidified melt) interface which are triggered by solutal effects as well as heat extraction or absorption due to phase changes occurring at microscale (explained in subsequent sections). CET is dominant when thermal gradient is small.

4.3.3 EVOLUTION OF PROBABILISTIC MODELS

The solution to above four problems is presented first by Brown and Spittle (Brown and Spittle 1989; Spittle and Brown 1989). They developed probabilistic models. They used Monte Carlo (MC) procedure for explaining solidification phenomena developed in earlier research (Anderson et al. 1984). MC method is based upon minimizing of interfacial energy (which is practically calculated by using physical properties of material [Zr- and Fe-based BMGMC]) from literature and earlier published data or inference from extrapolation or interpolation of data as needed. Procedurally, this minimizations is achieved by

a. Considering the energy of "unlike sites," for example, (a) "liquid/ solid sites" or (b) "sites belonging to different grains" and
b. By allowing transition between these states to occur according to randomly generated numbers

By using this method, Brown and Spittle were merely able to produce computed 2D microstructures which resembled very closely to those observed in real micrographic cross section. In particular

a. The selection of grains in the columnar zone and
b. CET

were nicely reproduced using this technique; also

a. the effect of solute concentration or
b. melt superheat upon the resultant microstructure

was determined "qualitatively" in a nice way. Their quantitative representation was not achieved.

4.3.3.1 Limitations

These methods suffer consistently from lack of physical basis and thus cannot be used to analyze quantitatively the effect of various physical phenomena (happening within the phase transformations). For example, to illustrate this, consider the following example:

a. During one MC time step, consider N sites where N is the number of sites whose evolution is calculated and is chosen from another N (total number) sites. Therefore, not all sites of interest (i.e., those located near to solid–liquid interface) are investigated. This in turn leads to algorithm predicted grain competition in columnar region, which does not at all reflect the physical mechanisms observed in organic alloys.

b. Furthermore, the results are sensitive to type of MC network itself which is used for computations. Thus, a single powerful model is presented in present work which combines "advantages of probabilistic methods with those of deterministic approaches" to predict more accurately the grain structure in a casting.

4.3.4 TWO-DIMENSIONAL CA METHOD

For this purpose, for now, a 2D CA model is developed which is based upon physical mechanisms of nucleation and growth (NG) of dendritic grains. Its salient features are as follows:

1. Heterogeneous nucleation, which was modeled by means of a nucleation site distribution in deterministic solidification models, is treated in a similar way in present probabilistic approach.
2. If total density of grains which nucleate at a given undercooling is obtained from an average distribution (d_c = average (distribution)), the location of these sites is chosen randomly

$$d_c = \frac{a_1 + a_2 + a_3 + a_4 + a_5 + \ldots + a_n}{n} \qquad (4)$$

where $a_1, a_2, a_3, a_4, \ldots, a_n$ are distributions of grains 1, 2, 3 to n
where $n = R$ (R = real numbers)

3. Crystallographic orientation of a newly nucleated grain is also taken into account at random.
4. The growth kinetics of (a) dendrite tip and (b) side branches are also incorporated into the model in such a way that final simulated microstructure is independent of the "CA network" which is used for computations.

Although it produces micrographic cross sections very much similar to those already obtained by Brown and Spittle, the present model has a "sound physical basis" and can thus reflect effects of (a) cooling rate" or (b) "solute concentration" quantitatively.

4.3.4.1 Detailed Description

Physical background: Consider a BMGMC wedge shape casting as shown in Figure 4.2.

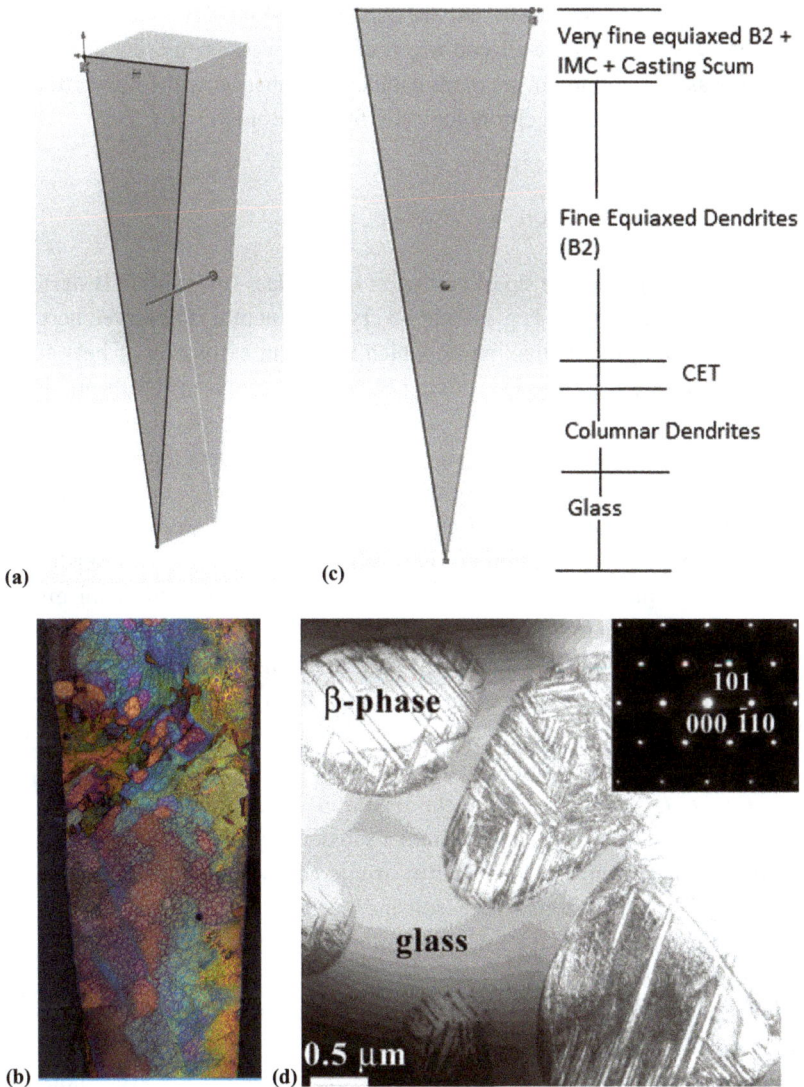

(a)

(c)

Very fine equiaxed B2 +
IMC + Casting Scum

Fine Equiaxed Dendrites
(B2)

CET

Columnar Dendrites

Glass

(b)

(d)

β-phase

glass

0.5 μm

$\overline{1}01$
000 $1\overline{1}0$

Figure 4.2 **(a)** 3D schematic; **(b)** optical micrograph of cross section (etched);
(c) 2D schematic showing regions; **(d)** a specific region (from B2 dendrites)
showing B19′ twins (B2–B19′ transformation-induced plasticity) (Pekarskaya
et al. 2001).

Figure 4.2(c) is a typical 2D cross section of cast eutectic Zr-based
BMGMC solidified in water-cooled wedge-shaped Cu mold (Gao et al.
2015; Gonzalez 2015). Their dendritic grains which have various crystallographic orientations appear as zones of different colors (Figure 4.2(b)).
Most common regions encountered in any casting appear here (Chalmers

1970; Kurz and Fisher 1986) and are marked all along the cross section. On the top end of wedge-shaped ingot, coarse grains are present as this region was exposed to air. Its more detailed explanation will follow after characterizing the region chronologically from bottom to top.

4.3.4.2 Characterization

Bottom region glass: The tip of casting is 100% glass (monolithic BMG). This region is classified as glass and no crystal structure is observed here because cooling rate is maximum, which results in extraction of heat at a very high rate resulting in retaining Supercooled liquid (SCL) state at room temperature.

Bottom region columnar dendrites: This region marks the beginning of "equiaxed columnar" first transition. This consists of very fine layer in which this transition happens and then columnar grains grow (primarily) in random 3D orientation because of still rapid rate of heat transfer which is complemented by sluggish nucleation on growth mechanisms of BMGMC. These grains are not very long as heat flow pattern is somewhat exponential because of wedge shape casting which triggers next transition too quickly before extension of growth as predicted by kinetics. This helps in retaining glassy matrix all throughout the casting. Otherwise 100% crystallization would have happened.

Bottom region CET: This is the region in which columnar dendritic grains which have developed/grown to a satisfactory level transit to equiaxed grains, known as CET. This is triggered by various phenomena such as solute diffusion, solute–solvent partitioning, shape of liquid–solid propagation front, thermal fluctuations happening at the tip of L–S propagating interface.

Fine equiaxed dendrites (B2): Once CET happens, equiaxed dendrites are formed all throughout the casting. Only their shape differs. In this region, they are fine sized while in *top region,* their size is even more reduced due to the presence of intermetallic compounds. Casting scum and other impurities couple with faster cooling rate from open top (convection and radiation) and side walls (conduction).

Note: In case of BMGMC not only inoculant particles serve as sites for heterogeneous nucleation but grain boundaries also serve this purpose (Porter and Easterling 1992; Song et al. 2016). Other defects and solidification microstructure also serve as sites for heterogeneous nucleation. (Their effects in total solidification [NG model] are to be taken into account in the final model.)

APPENDIX A

HETEROGENEOUS NUCLEATION AND GROWTH IN VERY FLUID ALLOYS (AS PER CLASSICAL NUCLEATION THEORY (CNT)) (HUNT 1984)

Heterogeneous nucleation rate per unit volume is defined as

$$I = N_s \upsilon \exp\left(-\frac{\Delta G_d}{kT}\right) \exp\left(-\frac{\Delta G_c}{kT}\right) \tag{1}$$

where

N_s = No. of atoms in contact with substrate

υ = vibrational frequency

ΔG_c = activation energy for nucleation (critical energy of nucleus formation, i.e., creation of liquid–solid interface)

ΔG_d = activation energy of diffusion (diffusional activation energy)

Rearranging equation (15) using definition of υ (vibrational frequency)

$$I = (N_o - N)I_o \exp\left(-\frac{\Delta G_c}{kT}\right)$$

$$I = (N_o I_o - NI_o) \exp\left(-\frac{\Delta G_c}{kT}\right) \tag{2}$$

Proof

$$I = N_o I_o e^{\left(\frac{\Delta G_c}{kT}\right)} - NI_o e^{\left(-\frac{\Delta G_c}{kT}\right)} \tag{3}$$

$$I = N_o \left(\frac{N_s}{t}\right) e^{\left(-\frac{\Delta G_c}{kT}\right)} - N\left(\frac{N_s}{t}\right) e^{\left(-\frac{\Delta G_c}{kT}\right)} \tag{4}$$

$$I = N_o \left(\frac{N_s}{t}\right) e^{\left(\frac{-\Delta G_c}{kT}\right)} - N\left(\frac{N_s}{t}\right) e^{\left(\frac{-\Delta G_c}{kT}\right)} \tag{5}$$

$$I = (\upsilon \times t)\left(\frac{N_s}{t}\right) e^{\left(\frac{-\Delta G_c}{kT}\right)} - (\upsilon \times t)\left(\frac{N_s}{t}\right) e^{\left(\frac{-\Delta G_c}{kT}\right)} \tag{6}$$

$$I = (N_s \times \upsilon) \, e^{\left(\frac{-\Delta G_c}{kT}\right)} - (N \times \upsilon) e^{\left(\frac{-\Delta G_c}{kT}\right)} \tag{7}$$

$$I = \upsilon \, e^{\left(\frac{-\Delta G_c}{kT}\right)} (N_s - N) \tag{8}$$

$$I = \upsilon \, e^{\left(\frac{-\Delta G_c}{kT}\right)} N_s \tag{9}$$

N can be neglected as during initial stages there is no nucleation event.

According to CNT, a minimum energy value is needed to create a solid–liquid interface eventually leading to stable nuclei out of melt. This is known as "activation energy." This activation energy is the energy to overcome ΔG^*—the energy barrier to nucleation. Now, as solid–liquid interface grows to form stable nuclei, atoms must be transported through liquid; thus, another temperature-dependent activation energy must be overcome known as ΔG_d (activation energy for diffusion).

The net effect is that CNT predicts a nucleation rate (I) given by

$$I = I_o \exp\left[-\frac{\left(\Delta G^* + \Delta G_d\right)}{k_\beta T} \right]$$

It is the nature of difference between ΔG^* and ΔG_d that dictates whether solidification will be crystalline or glassy. For crystalline solids, ΔG_d has a significant value, while for glassy solids there is no diffusion, thus ΔG_d can be neglected. Thus,

$$I = I_o \exp\left[-\frac{\Delta G^*}{k_\beta T} \right]$$

where k_β is a constant dictated by the nature and type of liquid composition and measured experimentally. ΔG_d is also zero in case of small undercooling (i.e., well-inoculated liquids/multicomponent alloys—metallic glasses inoculated with/without potent nuclei [present research]) (Browne et al. 2009).

Notes:

1. **Vibrational frequency**

$$\frac{N_s}{t} = \upsilon = \frac{\text{Total number of heterogeneous substrate particles}}{\text{Total time}}$$

$$\upsilon \times N_s$$

$$\frac{N_o}{t} \times N_s$$

where $N_s = I_o \times t$

or

$$I_o = \frac{N_s}{t}$$

That is,

Initial nucleation rate

$$= \frac{\text{Total number of atoms in contact with substrate}}{\text{Total time}}$$

Definition used in eq. (3).

2. The difference between frequency and rate is that frequency is (how many times an event happens in a given time?) while rate is (total number of that event (in terms of numerical value) per unit time). Thus, from equation (2),

N_o = total number of heterogeneous substrate particles originally available per unit volume

N = number that have already nucleated

I_o = constant

Value of I_o can be calculated from equation (15) using another term known as *liquid diffusion coefficient*.

$$D \approx a^2 \times \upsilon \exp\left(-\frac{\Delta G_d}{kT}\right) \tag{10}$$

where

$$a = \text{atomic diameter} = 0.4 \text{ nm}$$

$$\upsilon = \text{frequency}$$

which gives

$$I_o = 10^{18} - 10^{22}\!\big/_{\!s}$$

for small values of undercooling (well-inoculated melts/multicomponent alloys)

$$\Delta G_c \propto \frac{1}{(\Delta T)^2}$$

where ΔT = undercooling

Thus, nucleation rate is equation (2):

$$I = [N_o 10^{20} - N 10^{20}]\exp\left(-\frac{1}{kT(\Delta T)^2}\right)$$

$$\text{or } I = \left[N_o - N\right]10^{20}\exp\left(-\frac{u}{(\Delta T)^2}\right) \tag{11}$$

where u is a constant

$$u = \frac{1}{kT}$$

The value of u can be measured from
Method 1: T (heterogeneous nucleation temperature). This is defined as temperature, where there is an initial nucleation rate of 1 nucleus/cm³/sec.
Method 2: Second method to calculate u is

$$u = -\left(\Delta T_N\right)^2 \ln(N_o \times 10^{20})$$

Proof
Taking natural log of equation (25) on both sides

$$\ln(I) = \ln(N_o - N)10^{20}\left(-\frac{u}{(\Delta T_N)^2}\right)$$

$$-u = \left(\Delta T_N\right)^2\left[\ln I - \left[\ln(N_0 - N)10^{20}\right]\right]$$

$$-u = \left(\Delta T_N\right)^2\left[\ln I - \left[\ln N_0 10^{20} - \ln N 10^{20}\right]\right]$$

$$-u = \left(\Delta T_N\right)^2\left[\ln I - \ln N_0 10^{20} + \ln N 10^{20}\right]$$

because $\ln I$ and $\ln N10^{20}$ can be neglected

$$u = -\left(\Delta T_N\right)^2 \ln I - \ln(N_0 \times 10^{20})$$

where ΔT_N = undercooling at heterogeneous nucleation temperature.

Time is user-defined input and temperature comes from user-defined value initially as well. Then its every new value is assigned back to equation (1). With temperature and time, k changes and assigned back to equation (11). Also, with time, v (vibrational frequency) changes and is assigned back to original equation (1). Similarly, the value of u also changes with time and temperature. Table A.1 summarizes the values which are user defined and which change as a function of transience as program runs.

Notes:

1. In BMG, in some cases due to slow motion of large atoms, only nucleation happens and growth never happens. In these cases, a new phenomenon known as soft impingement effects of crystals must be considered. These could be solutal/thermal. However, this is quite rare.
2. In general, in case of BMG, CNT cannot be applied alone to describe NG.
3. Constitutional Supercooled Zone (CSZ) and interdependence models cannot be applied because of very high (η) viscosity of BMG (and their sluggish nature). CSZ and interdependence theories are for less viscous/more fluid alloys.
4. A new concept, known as complex interdiffusion tensor (Browne et al. 2009), is much more helpful to describe NG in BMG.
5. Fick's law in its native form (i.e., linear form) is not sufficient.

Table A.1. Summary of user-defined and program-determined functions used in CNT modified for BMGMC

Sr. No.	User defined value	Time	Temp$_{(f)}$
1	Time		
2	Temp$_{(i)}$		
3	Temp$_{(f)}$	✓	
4	K	✓	✓
5	Y	✓	✓
6	U	✓	✓

APPENDIX B

SPECIAL CASE OF GROWTH OF "COLUMNAR MICROSTRUCTURES"

The growth of columnar dendrites, which is initiated by nuclei that form at the mold interface (Cu mold casting/twin roll casting of bulk metallic glass matrix composite [BMGMC]) (only if CSZ is suppressed—not the present case), is usually simulated in a much simpler way. Again, in this case, there is no need to use cooling curve measurements or grain size measurement but same growth kinetics models (Kui et al. 1984; Wang et al. 2004) can be used.

Undercooling of eutectic front (ΔT_n eutectic) or

Undercooling of dendrite tips (ΔT_n dendrite tip) as well as

Undercooling of lamellae or dendrite trunk spacing (ΔT_n lamellae/trunk spacing)

This undercooling is determined by direct measurement of

a. Thermal gradient and
b. Speed of corresponding isotherm (eutectic or liquidus, respectively, i.e., speed of eutectic isotherm and speed of liquidus isotherm)

The latter values are obtained from macroscopic (part-scale) heat flow calculations (Rappaz 1989; Rafique and Iqbal 2009). The secondary arm spacing of both equiaxed and columnar dendritic microstructures is deduced from a local solidification time.

COLUMNAR STRUCTURE GROWTH IN WELL-INOCULATED BMGMC

Growth of columnar dendrites can also occur at the surface of external inoculants (well-inoculated deeply undercooled melts—present case of BMGMC development). However, it should also be noted that another condition for growth of columnar dendrite to occur is suppression of CSZ which clashes with aforementioned condition for onset of this phenomena at external potent nuclei. That is why there still is dispute about application of this concept to deeply undercooled well-inoculated melts (BMGMC) whose solution is under investigation.

COLUMNAR TO EQUIAXED TRANSITION (CET) (BASAK et al. 2016)

Growth rate of solid–liquid interface

$$V = S\cos\theta \qquad (1)$$

where S = scan speed.

Temperature gradient parallel to dendrite growth direction can be calculated using

$$G_{hkl} = G/\cos\psi \qquad (2)$$

where ψ = angle between "normal vector" and "possible dendrite growth orientation" at the solid–liquid interface. This is evaluated by CFX-post in Ansys.

A modification known as Rappaz modification is applied to predict CET. This is as follows:

$$\frac{G_{hkl}^n}{V_{hkl}} \geq a\left[3\sqrt{\frac{-4\pi}{3\ln(1-\phi)}}\sqrt{\frac{N_o}{n+1}\left(1 - \frac{\Delta T_n^{n+1}}{\Delta T_{tip}^{n+1}}\right)}\right]\exp(n) \qquad (3)$$

where V_{hkl} = dendrite growth velocity = $S\cos\theta / \cos\psi$,

    n = material constant determined from literature (Welk et al. 2014; Acharya and Das 2015),

    ϕ = equiaxed fraction (critical value = 0.066%),

    N_o = nucleation density,

    ΔT_{tip} = tip undercooling,

    ΔT_n = nucleation undercooling.

This will be incorporated in the present model at point where CET is determined. However, this model does not give true 3D representation output.

Note: In general, phase field methods are for microstructure evolution (its type [planer front, spherical front], morphology [precipitates, dendrites, and spheroids]), while cellular automation methods are for grain size determination (equiaxed/columnar dendritic) and its prediction. If both are combined (Tan et al. 2011; Choudhury et al. 2012; Zaeem et al. 2012), it is possible to get full map of microstructure evolution and grain size.

COMPARISON

Below a comparison of "strengths and capabilities" and "evolution of different theories over time" which have enabled a better understanding of nucleation and growth (NG) phenomena in bulk metallic glass (BMG) matrix composites (BMGMCs) is tabulated. The aim is to present the reader with a concise smart workable data for first-hand use and reference for solving NG problems in BMGMCs by modeling and simulation. This will help professional programmers, working engineers, and researchers to effectively find previously done research till now with its strengths and capabilities at one platform.

STRENGTHS AND CAPABILITIES

Table 1 compares strength, capabilities, and shortcomings of both deterministic and probabilistic methods. It highlights and chalks out parameters and certain segments of each technique which could possibly advantageously be used over others for modeling and simulation of BMGMCs. Note: N/A is an abbreviation to "Not Applicable."

Table 1. Comparison of strength and capabilities of modeling and simulation techniques as applied to NG problem of BMGMCs

Sr. No.	Phenomena/Property	Deterministic models		Probabilistic models		Comments	References
		Ductile phase	Glass	Ductile phase	Glass		
1	Nucleation (heterogeneous)	✓	N/A	✓	N/A		Rappaz (1989)
2	Growth	✓	✓	✓	✓		Hunt (1984) and Browne et al. (2009)
3	Growth mechanism (interdependence theory/complex interdiffusion tensor)	✓	N/A	✗	✗		Browne et al. (2009) and St John et al. (2011)
4	Different types of undercoolings (M-H model)	✓	✓	N/A	N/A		Maxwell and Hellawell (1975)
5	Growth kinetics	✗	✗	✓	✓		Rappaz and Gandin (1993)
6	Velocity of growth	✓	✓	✓	✓		Rappaz and Gandin (1993), Kurz et al. (1986), and Trivedi et al. (1987)
7	CET	✓	N/A	✓	N/A	Deterministic models can model ductile phase in 2D only	Rappaz and Gandin (1993), Basak et al. (2016), and Acharya et al. (2012)

#	Feature				Description	References
8	Impingement after CET	✓	N/A	N/A	Deterministic models can model ductile phase by JMAK correction, geometrical and random grain arrangement models only	Rappaz and Gandin (1993), Christian (2002), Avrami (1939), Price (1987), and Zou (1989)
9	Grain selection — Qualitatively	✗	✗	N/A	Probabilistic models can model ductile phase by MC only	Rappaz and Gandin (1993)
	Quantitatively		✓		Probabilistic models can model ductile phase by CA only	
10	Columnar dendrite arm branching	✗	✗	✓		Rappaz and Gandin (1993) and Gandin and Rappaz (1994)
11	Extension of grain	✗	✗	✓		Charbon and Rappaz (1993) and Gandin and Rappaz (1994)
12	CET in 3D	✗	N/A	N/A		Charbon and Rappaz (1993), Gandin and Rappaz (1997), and Gandin et al. (1999)

(continued)

Table 1. (*continued*)

Sr. No.	Phenomena/Property	Deterministic models		Probabilistic models		Comments	References
		Ductile phase	Glass	Ductile phase	Glass		
13	Physical basis	N/A	N/A	✓	✓	Probabilistic models can form the basis of modeling by MC	Rappaz and Gandin (1993) and Gandin and Rappaz (1994)
				✓	✓	Probabilistic models form the basis of modeling by CA	
14	Quantitative			✓	✓	Probabilistic models can only model quantitatively employing CA method	Rappaz and Gandin (1993) and Gandin and Rappaz (1994)
15	Liquid–liquid transition	✓	✓	N/A	N/A		Lan et al. (2016), Zu (2015), Wei et al. (2013)
16	Devitrification	✓	✓	✓	✓	Probabilistic models can model ductile and glass phase by 2D CA	Kelton (1998) and Blázquez (2008)

EVOLUTION OF THEORIES

Table 2. Evolution of theories of modeling and simulation as applied to NG problem of BMGMCs

Sr. No.	Method/Theory/Approach	Action and explanation	Limitation to explanation	Group/ Institute	Year	Reference
		Part-Scale Modeling				
		Analytical Modeling				
1	Deterministic/continuum model	Nonrandom methods, produce same types of exact results	Does not depend on initial state/point		1993	Gránásy (1993)
2	Probabilistic/stochastic models	Randomized result-based methods	Does depend on initial state		2016	Guo et al. (2016) and Nandi et al. (2016)
		Computational Modeling				
3	Lattice Boltzmann methods	Solution of basic "continuity" and "Navier–Stokes" equations for computational fluid dynamics (CFD) based on Ludwig Boltzmann's kinetic theory of gases	Limited to CFD-type problems	Raabe, D. (MPIE, Dusseldorf)	2004	Bao and Meskas (2011), Raabe (2004), Sun et al. (2009, 2011a, 2011b), Eshraghi, Jelinek, and Felicelli (2015), and Asle Zaeem (2015)

(continued)

Table 2. (*continued*)

Sr. No.	Method/Theory/Approach	Action and explanation	Limitation to explanation	Group/ Institute	Year	Reference
4	Phase field method	Solution of phase field (PF) parameter ϕ to describe physical state (liquid/solid) of material	Limited by type of ϕ for a particular situation	Napolitano, R. E. (Iowa State)	2002 and 2012	Boettinger et al. (2002) and Wang and Napolitano (2012)
5	Cellular automaton (CA) method	Division of entire volume into finite cells and solution of transport equations applied to individual cell	Large initial capital (processor/RAM)	Rappaz, M. (EPFL)	1993	Rappaz and Gandin (1993), Charbon and Rappaz (1993), Zhang et al. (2012), and Wang et al. (2016a, 2016b)
6	Virtual front tracking method	Dendritic growth in low Péclet number systems	Best in 2D	Stefanescu, D. M. (OSU)	2007	Zhu and Stefanescu (2007)
7	Sharp interface method	Evolution of interface as a function of time	Best in simple cases	Vermolen, F. J. (Delft)	2006	Vermolen (2006)
8	CAFÉ	Combine CA scheme with finite element (FE) method		Rappaz, M. (EPFL)	1994	Gandin and Rappaz (1994)

9	PFFE	Combine PF with FE method		Britta Nestler (KIT)	2011	Nestler and Choudhury (2011)
10	PFCA	Combine PF with CA regime		Shin, Y. C. (Purdue)	2011	Tan et al. (2011)
Atomistic Modeling						
10	Classical MD	Exact solutions	Computing power	Alder and Wainwright (Lawrence Livermore)	1957	Amokrane et al. (2015), Wang and Xu (2001), Alder and Wainwright (1957)
11	Monte Carlo simulation	Set of probability-based possible outcomes	Range of solutions	Metropolis, Nicholas, and coworkers (LANL)	1953 and 1993	Metropolis et al. (1953), Wood and Jacobson (1957), Pusztai and Sváb (1993), and Hwang et al. (2012)
12	Ab initio method/first principle calculation	Based on solution of Schrödinger equation	Works well for H atom only. For all other atoms, approximations are needed	Robert Parr (Caltech)	1950	Parr et al. (1950)

(continued)

Table 2. (*continued*)

Sr. No.	Method/Theory/Approach		Action and explanation	Limitation to explanation	Group/ Institute	Year	Reference
13	Hartree–Fock method and Slater determinant		Uses the variational theorem (which is wavefunction-based approach using mean field approximation)	Approximate solution is obtained. It is a form of ab initio method			Gilbert (2007)
14	Evolution of Hartree–Fock method	Self-constrained field method	Evolution of HF method	Approximate solutions			Gilbert (2007)
		Møller–Plesset (MP) perturbation (MP 1)	Hamiltonian is divided into two parts $\hat{H} = \hat{H}_0 + \lambda\hat{V}$ and solved	ψ and energy are HF ψ and HF energy			Gilbert (2007)

No.	Name				Year	Reference
	MP 2	ψ remains same, energy is changed	ψ is treated by the help of summations			Gilbert (2007)
	Density Functional Theory	Energy of system is obtained from electron density	Approximation based		1996	Perim et al. (2016), Wang et al. (2015), and Zheng (2012)
15	Interatomic potential	Explain interaction of atoms in a system in terms of potentials	Limited by accuracy, transferability, and computational speed of system	Multi (Many) body potentials Daw Baska (Sandia National Labs)	1984	Daw and Baskes (1984)

RESEARCH GAP

NG phenomena in single component (pure metals), binary, and multicomponent alloys are rather well understood. Classical nucleation theory (CNT) (Christian 2002) provides many answers to behavior of these melts. BMG and their composites (BMGMCs) are a relatively new class of materials which have recently emerged on the surface of science and technology and gained attention due to their unique properties (Chen 2011; Qiao 2013; Inoue 2015; Qiao et al. 2016). Traditionally, they were produced using conventional methods (Cu mold casting [Inoue and Zhang 1995; Inoue et al. 1998; Lou et al. 2011] and twin roll casting [TRC] [Inoue et al. 2015]) in which their metastable phase (glass) and any in situ ductile precipitates (stable phase) are nucleated based on their ability to surpass activation energy barrier. In addition, these processes impart very high cooling rate to castings which is essential for retention of supercooled liquid (glass) at room temperature explained by phenomena of confusion (Greer 1993), ordering (Greer 2006; Liu et al. 2010b; Ma 2015), frustration (Nelson 1983), and vitrification (Kumar et al. 2011; Wang 2012).

Very recently, with the advent and popularity of additive manufacturing (AM), interest has sparked to exploit the inherent and fundamental advantages present in this unique process to produce BMGs and BMGMCs. AM techniques are useful in achieving this objective as very high cooling rate in fusion liquid melt pool is already present inherently to assist the formation of glassy structure which is suppression of "kinetics" and prolonging of undercooling ("thermodynamics")—two main phenomena responsible for any phase transformation. However, the in situ nucleation of second phase equiaxed dendrites during solidification and then microstructural evolution (*solute diffusion* and *capillary* assisted) is not satisfactorily explained by CNT alone.

Either some modifications are needed in CNT or more reliable probabilistic microstructure evolution models (e.g., JMAK correction [Browne, Kovacs, and Mirihanage 2009]) are needed to explain NG (and other phenomena, e.g., liquid–liquid transition [James 1975; Zu 2015; Lan et al. 2016] and phase separations [Kim et al. 2013]) in BMGMCs. In this work, an effort has been made to meet both requirements. Following are propositions:

At the present scenario, there is no single hybrid/combined model which explains phenomena of heat transfer (liquid melt pool formation as a result of laser–matter interaction and its evolution—solidification) coupled with NG (solute diffusion [Wang and Beckermann 1993] and capillary action driven) at microscale to predict microstructure and grain size in BMGMC as melt cools in liquid pool of AM.

1. Only one study has been conducted to model the same phenomena (solidification only) during Cu mold suction casting which will serve as base (Li, Yan, and Wu 2009) in addition to very recent attempts (Browne, Kovacs, and Mirihanage 2009) in which emphasis is laid on development of generalized theory rather than solving a problem.

2. Only one study has been reported on microstructure formation during TRC using cellular automation finite element (CAFE; Wang et al. 2010a) but that is not aimed at BMGMCs, is carried out using commercial software package, and does not involve any mathematical modeling at the back-end. Software embedded (NG and heat transfer) models are only used.

3. Four prominent studies namely by Zhou et al. (2016), Zhang et al. (2013a), Zinoviev et al. (2016), and a group at Shenyang, China (Wang et al. 2014d; Tian et al. 2016) have been reported very recently using CAFE but these are based on modeling microstructure evolution in modified AM (HDMR [Zhou et al. 2016], LAMP [Chen et al. 2011] on 316L SS [Zhang et al. 2013a], 2D CAFE [Zinoviev et al. 2016], and cladding [Wang et al. 2014d; Tian et al. 2016]) processes.

4. Few studies in the past have been conducted employing selective laser melting using CAFE (Markl and Körner 2014; Wang et al. 2014d; Tian et al. 2016; Lindgren et al. 2016), CAPF, CAFVM (Fan et al. 2006c), modified CAFE (Zhu et al. 2004), etc., approaches, but none have been conducted on BMGMC.

5. No effort has been made to correlate the effect of edge-to-edge matching with assigning direction of easy heat flow and easy crystallographic growth.

6. No substantial study has been reported about evolution of microstructure in three dimensions in BMGMCs in AM.

7. No effort has been made to combine the effect of changing properties with decrease of temperature (transient conditions). Most of the models till now predict solutions in terms of steady-state processes.

8. Very few studies have been carried out to combine CA with FE in case of AM while it is routine approach to predict grain size in case of other processes (casting, welding).

OVERALL AIMS/RESEARCH QUESTIONS

Following research questions will be addressed in the present study:

1. How can potent inoculants be designed for controlling number density (d_c) and distribution of ductile phase in BMGMC via solidification?
2. What is the effect of number density (d_c), size, and distribution of ductile phase on final mechanical properties of BMGMCs?
3. What is the link between crystallographic matching model and heat flow in case of Zr-based BMGMCs?
4. For laser surface remelting process, what is the role of changing process variables (transient conditions—thermal conductivity of mold, temperature of melt, temperature of mold, viscosity of melt, etc.) on final microstructure evolution of BMGMC?
5. How do multiple thermal cycles affect the final microstructure of BMGMCs processed by AM?
6. How does coupling of modeling and simulation of microstructural evolution and heat transfer in BMGMC during AM using advanced programming and simulation platforms help in understanding NG phenomena?

METHODOLOGY

Following methodology will be adopted to realize the conceived model and its simulation:

Step 1: Write codes for transient conditions in MatLab using standard transport equations.
- Heat transfer equations to calculate time during each step
- Mass transfer (NG dominated by solute diffusion [Wang and Beckermann 1993] and capillary) equations to determine grain size during solidification steps only.

Step 2: Make part/section drawings representing liquid melt pool in AM in Ansys/SolidWorks
Import/integrate MatLab code
Determine mesh size and optimize it
Select properties (from literature and software libraries) and surfaces
Run simulation

Following scheme will be used to make simulate sets. Total 125 simulation sets will be run.

Sr. No	No. Density (dc)	Particle Size (micron)	Distribution
1	20	5	50
2			100
3			150
4			200
5			250

Sr. No	No. Density (dc)	Particle Size	Distribution
1	20	10	50
2			100
3			150
4			200
5			250

Sr. No	No. Density (dc)	Particle Size	Distribution
1	20	15	50
2			100
3			150
4			200
5			250

Sr. No	No. Density (dc)	Particle Size	Distribution
1	20	20	50
2			100
3			150
4			200
5			250

Sr. No	No. Density (dc)	Particle Size	Distribution
1	20	25	50
2			100
3			150
4			200
5			250

Sr. No	No. Density (dc)	Particle Size (micron)	Distribution
1	10	5	50
2			100
3			150
4			200
5			250

Sr. No	No. Density (dc)	Particle Size	Distribution
1	10	10	50
2			100
3			150
4			200
5			250

Sr. No	No. Density (dc)	Particle Size	Distribution
1	10	15	50
2			100
3			150
4			200
5			250

Sr. No	No. Density (dc)	Particle Size	Distribution
1	10	20	50
2			100
3			150
4			200
5			250

Sr. No	No. Density (dc)	Particle Size	Distribution
1	10	25	50
2			100
3			150
4			200
5			250

Total Number of Samples 125

Table 1

Sr. No	No. Density (dc)	Particle Size (micron)	Distribution
1	30	5	50
2			100
3			150
4			200
5			250

Table 2

Sr. No	No. Density (dc)	Particle Size	Distribution
1	30	10	50
2			100
3			150
4			200
5			250

Table 3

Sr. No	No. Density (dc)	Particle Size	Distribution
1	30	15	50
2			100
3			150
4			200
5			250

Table 4

Sr. No	No. Density (dc)	Particle Size	Distribution
1	30	20	50
2			100
3			150
4			200
5			250

Table 5

Sr. No	No. Density (dc)	Particle Size	Distribution
1	30	25	50
2			100
3			150
4			200
5			250

Table 6

Sr. No	No. Density (dc)	Particle Size (micron)	Distribution
1	40	5	50
2			100
3			150
4			200
5			250

Table 7

Sr. No	No. Density (dc)	Particle Size (micron)	Distribution
1	40	10	50
2			100
3			150
4			200
5			250

Table 8

Sr. No	No. Density (dc)	Particle Size (micron)	Distribution
1	40	15	50
2			100
3			150
4			200
5			250

Table 9

Sr. No	No. Density (dc)	Particle Size (micron)	Distribution
1	40	20	50
2			100
3			150
4			200
5			250

Table 10

Sr. No	No. Density (dc)	Particle Size (micron)	Distribution
1	40	25	50
2			100
3			150
4			200
5			250

Sr. No	No. Density (dc)	Particle Size (micron)	Distribution
1	50	5	50
2			100
3			150
4			200
5			250

Sr. No	No. Density (dc)	Particle Size (micron)	Distribution
1	50	10	50
2			100
3			150
4			200
5			250

Sr. No	No. Density (dc)	Particle Size (micron)	Distribution
1	50	15	50
2			100
3			150
4			200
5			250

Sr. No	No. Density (dc)	Particle Size (micron)	Distribution
1	50	20	50
2			100
3			150
4			200
5			250

Sr. No	No. Density (dc)	Particle Size (micron)	Distribution
1	50	25	50
2			100
3			150
4			200
5			250

Step 3: Experimental Validation of Microstructures

For verification of simulation results of microstructural development (modeling), following observation/experimental approach will be adopted.

OBSERVATION IN-SITU

High-speed camera to observe melt pool in AM and/or

Radiation pyrometer with imaging system and/or

Thermocouples (temperature measurement and its correlation with microstructure) (Different phases solidify at different cooling rate, which can give information about type of microstructure evolved during solidification.)

Optical micrography of actual castings (Cu mold wedge)/AM part will be carried out after appropriate etching (type, number density, size, and distribution will be compared with simulated results and correlations drawn).

DETERMINATION OF NUMBER DENSITY (d_c) FROM TTT DIAGRAM

Get TTT diagram of exact alloy composition (BMGMC) (Weinberg et al. 1989; Mukherjee et al. 2004) and determine critical cooling rate just enough to form (nucleate) a certain percentage of one crystalline phase in another (in the present case, glass). Then design apparatus accordingly. Objective is not to form monolithic glass but crystal phase embedded glass (i.e., ductile glass). Further, TTT diagram is used to calculate volume fraction of phases anticipated and these are compared with simulated and experimental results.

PROGRESS SO FAR

Ductile BMGMC samples have been casted in wedge shape using Cu mold suction casting at CSIRO, Clayton, Victoria. Two main compositions namely eutectic ($Zr_{47}Cu_{45.5}Al_5Co_2$) and other hypoeutectic ($Zr_{65}Cu_{15}Al_{10}Ni_{10}$) (described in detail in Section 1.1.13) are casted.

1. $Zr_{47}Cu_{45.5}Al_5Co_2$: Optical (Figure 1) and scanning electron microscopy (back-scattered electron [BSE] imaging [Figure 4]) have been performed on $Zr_{47}Cu_{45.5}Al_5Co_2$ after etching with solution of HF and HNO_3 in water.
2. Scanning electron microscopy (secondary electron [SE] imaging [Figure 2]) has been performed on $Zr_{65}Cu_{15}Al_{10}Ni_{10}$ after light etching with HNO_3.

The micrographs taken are shown in below figures and they clearly indicate presence of ductile B2 phase in the form of equiaxed dendrites originating from liquid upon solidification (Figures 1 and 4). Furthermore, etched surfaces of BMGMCs show pitting (Figures 2 and 3), which is a characteristic feature observed in this class of alloys upon treatment with HF acid.

These features, their appearance and effect of number density, size, and distribution of ductile phase, will be studied with modeling and simulation during upcoming 2 years to develop a deeper and better understanding of nucleation phenomena in BMGMCs.

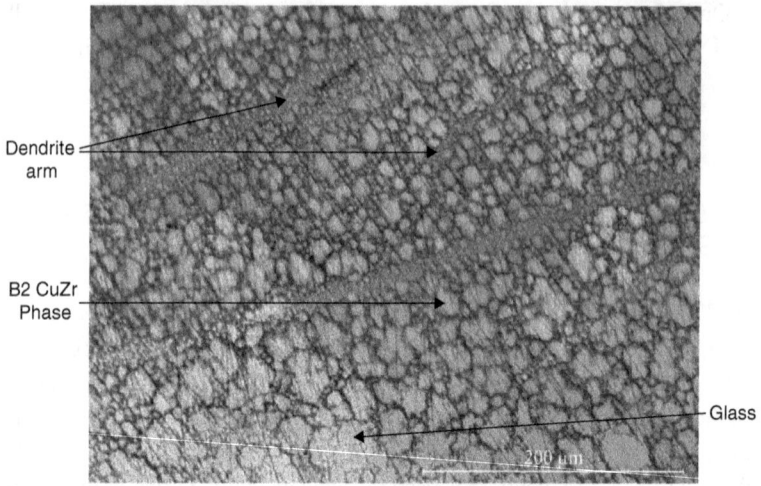

Figure 1. Incident light optical micrographs (without polarizer) of cross section of wedge sample of $Zr_{47}Cu_{45.5}Al_5Co_2$ (etched: HF + HNO_3 + H_2O).

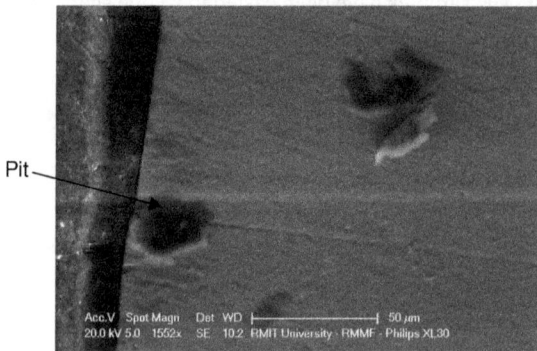

Figure 2. Secondary electron image of cross section of wedge sample of $Zr_{65}Cu_{15}Al_{10}Ni_{10}$ (light etch: HNO_3 + H_2O).

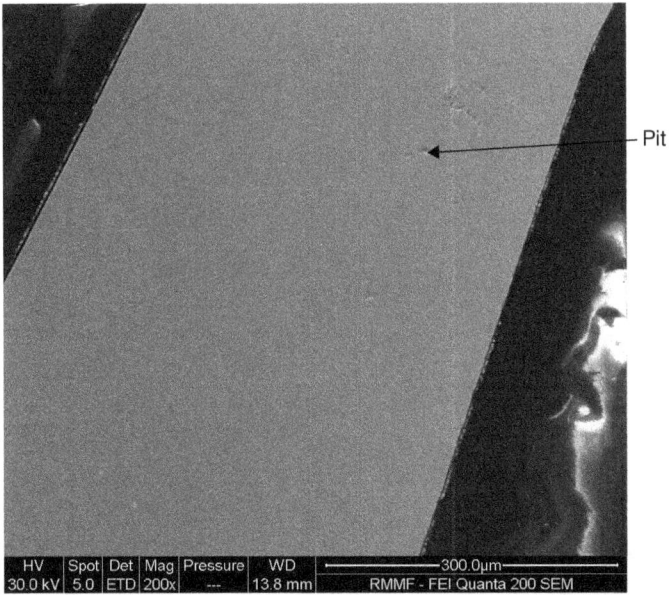

Figure 3. Low-magnification SE image of cross section of wedge sample of $Zr_{47}Cu_{45.5}Al_5Co_2$ (etched: $HF + HNO_3 + H_2O$) (showing pits) in glassy matrix.

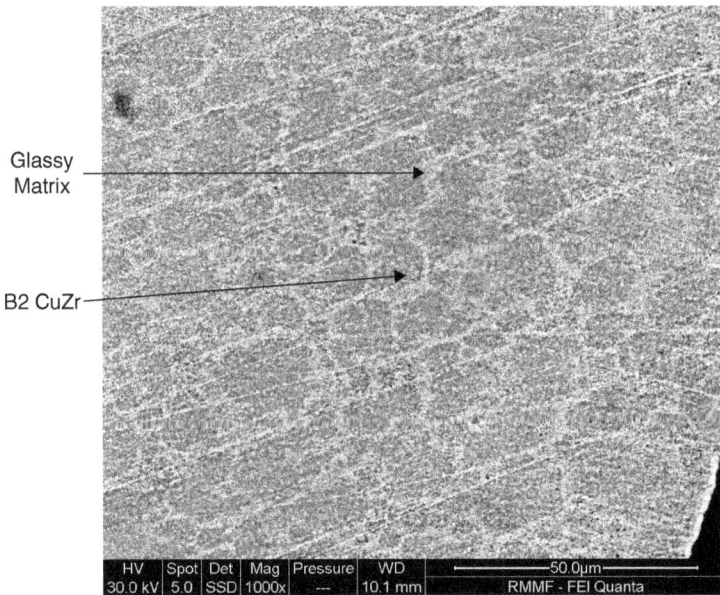

Figure 4. BSE image of cross section of wedge sample of $Zr_{47}Cu_{45.5}Al_5Co_2$ (etched: $HF + HNO_3 + H_2O$) (clearly showing dendrites, B2 phase, interdendritic network, and glassy matrix).

References

Abdeljawad, F., M. Fontus, and M. Haataja. 2011. "Ductility of Bulk Metallic Glass Composites: Microstructural Effects." *Applied Physics Letters* 98, no. 3, p. 031909.

Ablitzer, D. 1993. "Transport Phenomena and Modelling in Melting and Refining Processes." *Journal de Physique IV France* 03, no. C7, pp. C7-873–82.

Acharya, R., R. Bansal, J.J. Gambone, and S. Das. 2013. "Computational Modeling and Experimental Validation of Microstructural Development in Superalloy CMSX-4 Processed Through Scanning Laser Epitaxy." In *Solid Freeform Fabrication Symposium*. https://link.springer.com/chapter/10.1007/978-3-319-48764-9_336#citeas.

Acharya, R. and S. Das. 2015. "Additive Manufacturing of IN100 Superalloy Through Scanning Laser Epitaxy for Turbine Engine Hot–Section Component Repair: Process Development, Modeling, Microstructural Characterization, and Process Control." *Metallurgical and Materials Transactions A* 46, no. 9, pp. 3864–75.

Ågren, J. 1982. "Numerical Treatment of Diffusional Reactions in Multicomponent Alloys." *Journal of Physics and Chemistry of Solids* 43, no. 4, pp. 385–91.

Akhtar, D., B. Cantor, and R.W. Cahn. 1982a. "Diffusion Rates of Metals in a NiZr2 Metallic Glass." *Scripta Metallurgica* 16, no. 4, pp. 417–20.

Akhtar, D., B. Cantor, and R.W. Cahn. 1982b. "Measurements of Diffusion Rates of Au in Metal–Metal and Metal–Metalloid Glasses." *Acta Metallurgica* 30, no. 8, pp. 1571–7.

Akhtar, D. and R.D.K. Misra. 1985. "Impurity Diffusion in a Ni–Nb Metallic Glass." *Scripta Metallurgica* 19, no. 5, pp. 603–7.

Akhtar, D. and R.D.K. Misra. 1986. "Effect of Thermal Relaxation on Diffusion in a Metallic Glass." *Scripta Metallurgica* 20, no. 5, pp. 627–31.

Akihisa, I., K. Ohtera, A.-P. Tsai, and T. Masumoto. 1988. "Aluminum-Based Amorphous Alloys with Tensile Strength above 980 MPa (100 kg/mm^2)." *Japanese Journal of Applied Physics* 27, no. 4A, p. L479.

Alder, B.J. and T.E. Wainwright. 1957. "Phase Transition for a Hard Sphere System." *The Journal of Chemical Physics* 27, no. 5, pp. 1208–9.

Amine, T., J.W. Newkirk, and F. Liou. 2014a. "An Investigation of the Effect of Direct Metal Deposition Parameters on the Characteristics of the Deposited Layers." *Case Studies in Thermal Engineering* 3, pp. 21–34.

Amine, T., J.W. Newkirk, and F. Liou. 2014b. "Investigation of Effect of Process Parameters on Multilayer Builds by Direct Metal Deposition." *Applied Thermal Engineering* 73, no. 1, pp. 500–11.

Amokrane, S., A. Ayadim, and L. Levrel. 2015. "Structure of the Glass–Forming Metallic Liquids by Ab-Initio and Classical Molecular Dynamics, A Case Study: Quenching the Cu60Ti20Zr20 Alloy." *Journal of Applied Physics* 118, no. 19, p. 194903.

An, Z.N., W.D. Li, F.X. Liu, P.K. Liaw, and Y.F. Gao. 2012. "Interface Constraints on Shear Band Patterns in Bonded Metallic Glass Films Under Microindentation." *Metallurgical and Materials Transactions A* 43, no. 8, pp. 2729–41.

Anderson, M., D.J. Srolovitz, G.S. Grest, and P.S. Sahni. 1984. "Computer Simulation of Grain Growth—I. Kinetics." *Acta Metallurgica* 32, no. 5, pp. 783–91.

Antonaglia, J., W.J. Wright, X. Gu, G.G. Byer, T.C. Hufnagel, M. LeBlanc, J.T. Uhl, and K.A. Dahmen. 2014a. "Bulk Metallic Glasses Deform via Slip Avalanches." *Physical Review Letters* 112, no. 15, p. 155501.

Antonaglia, J., X. Xie, G. Schwarz, M. Wriath, J. Qiao, Y. Zhang, P.K. Liaw, T. Uhl, and K.A. Dahmen. 2014b. "Tuned Critical Avalanche Scaling in Bulk Metallic Glasses." *Scientific Reports* 4, p. 4382.

Antonie, O. 2007. *Langevin Dynamics of Protein Folding: Influence of Confinement* [Masters Degree Project]. Scuola Internazionale Superiore di Studi Avanzati, SISSA, Trieste, Italy.

Antonione, C., S. Sprianoa, P. Rizzib, M. Baricoob, and L. Battezzatib. 1998. "Phase Separation in Multicomponent Amorphous Alloys." *Journal of Non-Crystalline Solids* 232–234, pp. 127–32.

Antonowicz, J., E. Jezierska, M. Kedzierski, A.R. Yavari, A.L. Green, P. Panine, and M. Sztucki. 2008. "Early Stages of Phase Separation and Nanocrystallization in Al-Rare Earth Metallic Glasses Studied Using SAXS/WAXS and HRTEM Methods." *Reviews on Advanced Materials Science* 18, no. 5, pp. 454–8.

Argon, A.S. 1979. "Plastic Deformation in Metallic Glasses." *Acta Metallurgica* 27, no. 1, pp. 47–58.

Asle Zaeem, M. 2015. "Advances in Modeling of Solidification Microstructures." *JOM* 67, no. 8, pp. 1774–5.

Atwood, C., M. Ensz, D. Greene, M. Griffith, L. Harwell, D. Reckaway, T. Romero, E. Schlienger, and J. Smugeresky. 1998. Laser engineered net shaping (LENS): A tool for direct fabrication of metal parts. In *Proceedings of International Congress on Applications of Lasers and Electro – Optics (ICALEO)*, Orlando, FL.

Avrami, M. 1939. "Kinetics of Phase Change. I. General Theory." *The Journal of Chemical Physics* 7, no. 12, pp. 1103–12.

Badrossamay, M. and T.H.C. Childs. 2007. "Further Studies in Selective Laser Melting of Stainless and Tool Steel Powders." *International Journal of Machine Tools and Manufacture* 47, no. 5, pp. 779–84.

Bai, L., B. Wang, H. Zhong, J. Ni, Q. Zhai, and J. Zhang. 2016. "Experimental and Numerical Simulations of the Solidification Process in Continuous Casting of Slab." *Metals* 6, no. 3, p. 53.

Balla, V.K. and A. Bandyopadhyay. 2010. "Laser Processing of Fe-based Bulk Amorphous Alloy." *Surface and Coatings Technology* 205, no. 7, pp. 2661–7.

Balla, V.K., S. Banerjee, S. Bose, and A. Bandyopadhyay. 2010. "Direct Laser Processing of a Tantalum Coating on Titanium for Bone Replacement Structures." *Acta Biomaterialia* 6, no. 6, pp. 2329–34.

Balla, V.K., W. Xue, S. Bose, and A. Bandyopadhyay. 2009. "Laser-Assisted Zr/ZrO2 Coating on Ti for Load-Bearing Implants." *Acta Biomaterialia* 5, no. 7, pp. 2800–9.

Bao, Y.B. and J. Meskas. 2011. *Lattice Boltzmann Method for Fluid Simulations.* Department of Mathematics, Courant Institute of Mathematical Sciences, New York University.

Baricco, M., Palumbo, M., D. Baldissin, E. Bose, and L. Battezatti. 2004. "Metastable Phases and Phase Diagrams." *La Metallurgia Italiana* 3, 1–8.

Basak, A., R. Acharya, and S. Das. 2016. "Additive Manufacturing of Single–Crystal Superalloy CMSX–4 Through Scanning Laser Epitaxy: Computational Modeling, Experimental Process Development, and Process Parameter Optimization." *Metallurgical and Materials Transactions A* 47, no. 8, pp. 3845–59.

Baskes, M.I. 1992. "Modified Embedded-Atom Potentials for Cubic Materials and Impurities." *Physical Review B* 46, no. 5, pp. 2727–42.

Basu, J., N. Najendra, Y. Li, and U. Ramamurty. 2003. "Microstructure and Mechanical Properties of a Partially Crystallized La-based Bulk Metallic Glass." *Philosophical Magazine* 83, no. 15, pp. 1747–60.

Basu, A., A.N. Samant, S.P. Harimkar, D.J. Majumdar, I. Manna, and N.B. Dahotre. 2008. "Laser Surface Coating of Fe–Cr–Mo–Y–B–C Bulk Metallic Glass Composition on AISI 4140 Steel." *Surface and Coatings Technology* 202, no. 12, pp. 2623–31.

Baufeld, B., E. Brandl, and O. van der Biest. 2011. "Wire Based Additive Layer Manufacturing: Comparison of Microstructure and Mechanical Properties of Ti–6Al–4V Components Fabricated by Laser-Beam Deposition and Shaped Metal Deposition." *Journal of Materials Processing Technology* 211, no. 6, pp. 1146–58.

Bennon, W.D. and F.P. Incropera. 1987. "A Continuum Model for Momentum, Heat and Species Transport in Binary Solid-Liquid Phase Change Systems—I. Model Formulation." *International Journal of Heat and Mass Transfer* 30, no. 10, pp. 2161–70.

Bergmann, H.-W. and B.L. Mordike. 1981. "Laser and Electron-Beam Melted Amorphous Layers." *Journal of Materials Science* 16, no. 4, pp. 863–9.

Berzins, M., T.H.C. Childs, and G.R. Ryder. 1996. "The Selective Laser Sintering of Polycarbonate." *CIRP Annals - Manufacturing Technology* 45, no. 1, pp. 187–90.

Bhanumurthy, K., G.K. Dey, and S. Banerjee. 1988. "Metallic Glass Formation by Solid State Reaction in Bulk Zirconium – Copper Diffusion Couples." *Scripta Metallurgica* 22, no. 9, pp. 1395–8.

Biffi, C.A., A. Figini Albisetti, and A. Tuissi. 2013. "CuZr Based Shape Memory Alloys: Effect of Cr and Co on the Martensitic Transformation." In *Materials Science Forum,* Trans Tech Publications, Switzerland.

Binder, K. and P. Virnau. 2016. "Overview: Understanding Nucleation Phenomena From Simulations of Lattice Gas Models." *The Journal of Chemical Physics* 145, no. 21, p. 211701.

Blázquez, J.S., V. Franco, C.F. Conde, M. Millan, and A. Conde. 2008. "Instantaneous Growth Approximation Describing the Nanocrystallization Process of Amorphous Alloys: A Cellular Automata Model." *Journal of Non-Crystalline Solids* 354, no. 30, pp. 3597–605.

Boettinger, W.J., J.A. Warren, C. Beckermann, and A. Karma. 2002. "Phase-Field Simulation of Solidification." *Annual Review of Materials Research* 32, no. 1, pp. 163–94.

Borkar, T., B. Gwalani, D. Choudhuri, C.V. Mikler, C.J. Yannetta, X. Chen, R.V. Ramanujan, M.J. Styles, M.A. Gibson, and R. Banerjee. 2016. "A Combinatorial Assessment of AlxCrCuFeNi2 ($0 < x < 1.5$) Complex Concentrated Alloys: Microstructure, Microhardness, and Magnetic Properties." *Acta Materialia* 116, pp. 63–76.

Böttger, B., J. Eiken, and I. Steinbach. 2006. "Phase Field Simulation of Equiaxed Solidification in Technical Alloys." *Acta Materialia* 54, no. 10, pp. 2697–704.

Bracchi, A., Y.-L. Huang, M. Seibt, and S. Schneider. 2006. "Decomposition and Metastable Phase Formation in the Bulk Metallic Glass Matrix Composite Zr56Ti14Nb5Cu7Ni6Be12." *Journal of Applied Physics* 99, no. 12, p. 123519.

Brazhkin, V.V. 2006a. "Metastable Phases and 'Metastable' Phase Diagrams." *Journal of Physics: Condensed Matter* 18, no. 42, p. 9643.

Brazhkin, V.V. 2006b. "Metastable Phases, Phase Transformations, and Phase Diagrams in Physics and Chemistry." *Physics-Uspekhi* 49, no. 7, pp. 719–24.

Brif, Y., M. Thomas, and I. Todd. 2015. "The Use of High-Entropy Alloys in Additive Manufacturing." *Scripta Materialia* 99, pp. 93–96.

Broecker, P. and S. Trebst. 2016. "Entanglement and the Fermion Sign Problem in Auxiliary Field Quantum Monte Carlo Simulations." *Physical Review B* 94, no. 7, p. 075144.

Brown, S.G.R. and J.A. Spittle. 1989. "Computer Simulation of Grain Growth and Macrostructure Development During Solidification." *Materials Science and Technology* 5, no. 4, pp. 362–8.

Browne, D.J., Z. Kovacs, and W.U. Mirihanage. 2009. "Comparison of Nucleation and Growth Mechanisms in Alloy Solidification to Those in Metallic Glass Crystallisation — Relevance to Modeling." *Transactions of the Indian Institute of Metals* 62, no. 4, pp. 409–12.

Buchbinder, D., H. Schleifenbaum, S. Heidrich, W. Meiners, and J. Bultmann. 2011. "High Power Selective Laser Melting (HP SLM) of Aluminum Parts." *Physics Procedia* 12, pp. 271–8.

Busetti, F. 2003. *Simulated Annealing Overview.* Italy: JP Morgan.

Caginalp, G. and W. Xie. 1993. "Phase-Field and Sharp-Interface Alloy Models." *Physical Review E* 48, no. 3, pp. 1897–909.

Chalmers, B. 1970. "Principles of Solidification." In *Applied Solid State Physics*, ed. W. Low and M. Schieber. Boston, MA: Springer. pp. 161–70.

Chang, H.J., W. Wook, E.S. Park, J.S. Kyeong, and D.H. Kim. 2010. "Synthesis of Metallic Glass Composites Using Phase Separation Phenomena." *Acta Materialia* 58, no. 7, pp. 2483–91.

Charbon, C. and M. Rappaz. 1993. "3D Probabilistic Modelling of Equiaxed Eutectic Solidification." *Modelling and Simulation in Materials Science and Engineering* 1, no. 4, p. 455.

Chen, B., S. Pang, P. Han, Y. Li, A.R. Yavari, A.R. Vaughan, and T. Zhang. 2010. "Improvement in Mechanical Properties of a Zr-Based Bulk Metallic Glass by

Laser Surface Treatment." *Journal of Alloys and Compounds* 504, Supplement 1, pp. S45–7.

Chen, C., M. Miller, and K. Sepehrnoori. 1997. A new 3D front-tracking approach to modeling displacement processes in complex large-scale reservoirs. In *SPE Reservoir Simulation Symposium,* Society of Petroleum Engineers.

Chen, C., Y. Xue, L. Wang, X. Cheng, F. Wang, Z. Wang, H. Zhang, and A. Wang. 2012. "Effect of Temperature on the Dynamic Mechanical Behaviors of Zr-Based Metallic Glass Reinforced Porous Tungsten Matrix Composite." *Advanced Engineering Materials* 14, no. 7, pp. 439–44.

Chen, G., H. Bei, Y. Cao, A. Gali, C.T. Liu, and E.P. George. 2009. "Enhanced Plasticity in a Zr-Based Bulk Metallic Glass Composite With In Situ Formed Intermetallic Phases." *Applied Physics Letters* 95, no. 8, p. 081908.

Chen, H.S. 1973. "Plastic Flow in Metallic Glasses Under Compression." *Scripta Metallurgica* 7, no. 9, pp. 931–5.

Chen, H.S. 1974. "Thermodynamic Considerations on the Formation and Stability of Metallic Glasses." *Acta Metallurgica* 22, no. 12, pp. 1505–11.

Chen, H.S. 1976. "Ductile-Brittle Transition in Metallic Glasses." *Materials Science and Engineering* 26, no. 1, pp. 79–82.

Chen, H.S. 1980. "Glassy Metals." *Reports on Progress in Physics* 43, no. 4, p. 353.

Chen, L.Y., Z.D. Fu, G.Q. Zhang, X.P. Hao, Q.K. Jiang, X.D. Wang, Q.P. Cao, et al. 2008. "New Class of Plastic Bulk Metallic Glass." *Physical Review Letters* 100, no. 7, p. 075501.

Chen, M. 2008. "Mechanical Behavior of Metallic Glasses: Microscopic Understanding of Strength and Ductility." *Annual Review of Materials Research* 38, no. 1, pp. 445–69.

Chen, M. 2011. "A Brief Overview of Bulk Metallic Glasses." *NPG Asia Materials* 3, pp. 82–90.

Chen, S. 2014. *Three dimensional Cellular Automaton–Finite Element (CAFE) modeling for the grain structures development in Gas Tungsten/Metal Arc Welding processes.* Ecole Nationale Supérieure des Mines de Paris.

Chen, S., G. Guillemot, and C.-A. Gandin. 2014. "3D Coupled Cellular Automaton (CA)–Finite Element (FE) Modeling for Solidification Grain Structures in Gas Tungsten Arc Welding (GTAW)." *ISIJ International* 54, no. 2, pp. 401–7.

Chen, T.-H. and C.-K. Tsai. 2015. "The Microstructural Evolution and Mechanical Properties of Zr-Based Metallic Glass under Different Strain Rate Compressions." *Materials* 8, no. 4, p. 1831.

Chen, Y., A. Wang, H. Fu, Z. Zhu, H. Zhang, Z. Hu, L. Wang, and H. Chang. 2011. "Preparation, Microstructure and Deformation Behavior of Zr-Based Metallic Glass/Porous SiC Interpenetrating Phase Composites." *Materials Science and Engineering: A* 530, pp. 15–20.

Chen, Y., C. Zhou, and J. Lao. 2011. "A Layerless Additive Manufacturing Process Based on CNC Accumulation." *Rapid Prototyping Journal* 17, no. 3, pp. 218–27.

Cheng, J.L. and G. Chen. 2013. "Glass Formation of Zr–Cu–Ni–Al Bulk Metallic Glasses Correlated With L → Zr2Cu + ZrCu Pseudo Binary Eutectic Reaction." *Journal of Alloys and Compounds* 577, pp. 451–5.

Cheng, J.-L., G. Chen, C.T. Liu, and Y. Li. 2013. "Innovative Approach to the Design of Low-Cost Zr-Based BMG Composites With Good Glass Formation." *Scientific Reports* 3, p. 2097.

Cheng, Y.Q. and E. Ma. 2011. "Atomic-Level Structure and Structure–Property Relationship in Metallic Glasses." *Progress in Materials Science* 56, no. 4, pp. 379–473.

Cheng, Y.Q., E. Ma, and H.W. Sheng. 2008. "Alloying Strongly Influences the Structure, Dynamics, and Glass Forming Ability of Metallic Supercooled Liquids." *Applied Physics Letters* 93, no. 11, p. 111913.

Cheng, Y.Q., E. Ma, and H.W. Sheng. 2009. "Atomic Level Structure in Multicomponent Bulk Metallic Glass." *Physical Review Letters* 102, no. 24, p. 245501.

Childs, T., C. Hauser, C.M. Taylor, and A.E. Tontowi. 2000. "Simulation and experimental verification of crystalline polymer and direct metal selective laser sintering." In *Proceedings of SFF Symposium*, Austin, TX.

Choi-Yim, H. 1998. *Synthesis and Characterization of Bulk Metallic Glass Matrix Composites*. Pasadena, CA: California Institute of Technology.

Choi-Yim, H., R.D. Conner, F. Szuecs, and W.L. Johnson. 2001. "Quasistatic and Dynamic Deformation of Tungsten Reinforced Zr57Nb5Al10Cu15.4Ni12.6 Bulk Metallic Glass Matrix Composites." *Scripta Materialia* 45, no. 9, pp. 1039–45.

Choi-Yim, H., R.D. Conner, F. Szuecs, and W.L. Johnson. 2002. "Processing, Microstructure and Properties of Ductile Metal Particulate Reinforced Zr57Nb5Al10Cu15.4Ni12.6 Bulk Metallic Glass Composites." *Acta Materialia* 50, no. 10, pp. 2737–45.

Choi-Yim, H. and W.L. Johnson. 1997. "Bulk Metallic Glass Matrix Composites." *Applied Physics Letters* 71, no. 26, pp. 3808–10.

Choudhury, A., K. Reuther, E. Wesner, A. August, B. Nestler, and M. Rettenmayer. 2012. "Comparison of Phase-Field and Cellular Automaton Models for Dendritic Solidification in Al–Cu Alloy." *Computational Materials Science* 55, pp. 263–8.

Christian, J.W. 2002a. "Chapter 10 – The Classical Theory of Nucleation." In *The Theory of Transformations in Metals and Alloys*.Oxford: Pergamon, pp. 422–79.

Christian, J.W. 2002b. "Front Matter A2." In *The Theory of Transformations in Metals and Alloys*. Oxford: Pergamon, p. iii.

Chu, J.P. 2009. "Annealing-Induced Amorphization in a Glass-Forming Thin Film." *JOM* 61, no. 1, pp. 72–75.

Chu, J.P., J.S. Huang, C. Jang, Y.C. Wang, and P.K. Liaw. 2010. "Thin Film Metallic Glasses: Preparations, Properties, and Applications." *JOM* 62, no. 4, pp. 19–24.

Chu, J.P., J.S.C. Jang, J.C. Huang, H.S. Chou, Y. Yang, J.C. Ye, Y.C. Wang, et al. 2012. "Thin Film Metallic Glasses: Unique Properties and Potential Applications." *Thin Solid Films* 520, no. 16, pp. 5097–122.

Chu, M.Y., Z.M. Jiao, R.F. Wu, Z.H. Wang, H.J. Yang, Y.S. Wang, and J.W. Qiao. 2015. "Quasi-Static and Dynamic Deformation Behaviors of an In-Situ Ti-Based Metallic Glass Matrix Composite." *Journal of Alloys and Compounds* 640, pp. 305–10.

Conner, R.D., H. Choi-Yim, and W.L. Johnson. 1999. "Mechanical Properties of Zr57Nb5Al10Cu15.4Ni12.6 Metallic Glass Matrix Particulate Composites." *Journal of Materials Research* 14, no. 08, pp. 3292–7.

Conner, R.D., R.B. Dandliker, and W.L. Johnson. 1998. "Mechanical Properties of Tungsten and Steel Fiber Reinforced Zr41.25Ti13.75Cu12.5Ni10Be22.5 Metallic Glass Matrix Composites." *Acta Materialia* 46, no. 17, pp. 6089–102.

Conner, R.D., A.J. Rosakis, W.L. Johnson, and D.M. Owen. 1997. "Fracture Toughness Determination for a Beryllium-Bearing Bulk Metallic Glass." *Scripta Materialia* 37, no. 9, pp. 1373–8.

Cunliffe, A., J. Plummer, I. Figueroa, and I. Tod. 2012. "Glass Formation in a High Entropy Alloy System by Design." *Intermetallics* 23, pp. 204–7.

Cytron, S.J. 1982. "A Metallic Glass-Metal Matrix Composite." *Journal of Materials Science Letters* 1, no. 5, pp. 211–3.

Dandliker, R.B., R.D. Conner, and W.L. Johnson. 1998. "Melt Infiltration Casting of Bulk Metallic-Glass Matrix Composites." *Journal of Materials Research* 13, no. 10, pp. 2896–901.

Das, J., M. Bostrom, N. Mattern, A. Kvick, A.R. Yavari, A.L. Greer, and J. Eckert. 2007. "Plasticity in Bulk Metallic Glasses Investigated Via the Strain Distribution." *Physical Review B* 76, no. 9, p. 092203.

Das, J., K.B. Kim, and F. Baier, 2005a. "High-Strength Ti-Base Ultrafine Eutectic With Enhanced Ductility." *Applied Physics Letters* 87, no. 16, p. 161907.

Das, J., S. Pauly, M. Bostrom, K. Durst, M. Goken, and J. Eckert. 2009. "Designing Bulk Metallic Glass and Glass Matrix Composites in Martensitic Alloys." *Journal of Alloys and Compounds* 483, no. 1–2, pp. 97–101.

Das, J., M.B. Tang, K.B. Kim, R. Theissmann, F. Baier, W.H. Wang, and J. Eckert. 2005b. "'Work–Hardenable' Ductile Bulk Metallic Glass." *Physical Review Letters* 94, no. 20, p. 205501.

Daw, M.S. and M.I. Baskes. 1984. "Embedded-Atom Method: Derivation and Application to Impurities, Surfaces, and Other Defects in Metals." *Physical Review B* 29, no. 12, p. 6443.

Deckard, C.R. 1988. *Method and Apparatus for Producing Parts by Selective Sintering. Google Patents.*

Deng, S.T., H. Diao, Y.L. Chen, C. Yan, H.F. Zhang, A.M. Wang, and Z.Q. Hu. 2011. "Metallic Glass Fiber-Reinforced Zr-Based Bulk Metallic Glass." *Scripta Materialia* 64, no. 1, pp. 85–88.

Dijkhuizen, W., I. Roghair, M.V.S. Annaland, and J.A.M. Kuipers. 2010. "DNS of Gas Bubbles Behaviour Using an Improved 3D Front Tracking Model— Model Development." *Chemical Engineering Science* 65, no. 4, pp. 1427–37.

Ding, J., Z. Liu, H. Wang, and T. Zhang. 2014. "Large-Sized CuZr-Based Bulk Metallic Glass Composite With Enhanced Mechanical Properties." *Journal of Materials Science & Technology* 30, no. 6, pp. 590–4.

Ding, D., Z. Pan, D. Cuiuri, and H. Li. 2015. "Wire-Feed Additive Manufacturing of Metal Components: Technologies, Developments and Future Interests." *The International Journal of Advanced Manufacturing Technology* 81, no. 1, pp. 465–81.

Dodd, B. and Y. Bai. 2012. *Adiabatic Shear Localization: Frontiers and Advances.* London: Elsevier.

Donald, I.W. and H.A. Davies. 1978. "Prediction of Glass-Forming Ability for Metallic Systems." *Journal of Non-Crystalline Solids* 30, no. 1, pp. 77–85.

Donald, W.B., O.A. Shenderova, J.A. Harriso, S.J. Stuart, B. Ni, and S.B. Sinnott. 2002. "A Second-Generation Reactive Empirical Bond Order (REBO) Potential Energy Expression for Hydrocarbons." *Journal of Physics: Condensed Matter* 14, no. 4, p. 783.

Donovan, P.E. and W.M. Stobbs. 1981. "The Structure of Shear Bands in Metallic Glasses." *Acta Metallurgica* 29, no. 8, pp. 1419–36.

Donovan, P.E. and W.M. Stobbs. 1983. "Shear Band Interactions With Crystals in Partially Crystallised Metallic Glasses." *Journal of Non-Crystalline Solids* 55, no. 1, pp. 61–76.

Drehman, A.J., A.L. Greer, and D. Turnbull. 1982. "Bulk Formation of a Metallic Glass: Pd40Ni40P20." *Applied Physics Letters* 41, no. 8, pp. 716–7.

Drescher, P. and H. Seitz. 2015. "Processability of an Amorphous Metal Alloy Powder by Electron Beam Melting." *RTeJournal – Fachforum für Rapid Technologie* 2015, no. 1.

Du, X.H., J.C. Huang, K.C. Hsieh, Y.H. Lai, H.M. Chen, J.S.C. Jang, and P.K. Liaw. 2007. "Two-Glassy-Phase Bulk Metallic Glass With Remarkable Plasticity." *Applied Physics Letters* 91, no. 13, p. 131901.

Duan, G., A. Wiest, M.L. Lind, J. Li, W.K. Rhim, and W.L. Johnson. 2007. "Bulk Metallic Glass with Benchmark Thermoplastic Processability." *Advanced Materials* 19, no. 23, pp. 4272–5.

Dutta, B. and F.H. Froes. 2016. *Additive Manufacturing of Titanium Alloys.* Oxford: Butterworth–Heinemann, pp. 1–10.

Eckert, J., J. Das, S. Pauly, and C. Duhamel. 2007. "Mechanical Properties of Bulk Metallic Glasses and Composites." *Journal of Materials Research* 22, no. 02, pp. 285–301.

Eckert, J., U. Kuhn, N. Mattern, G. He, and A. Gebert. 2002. "Structural Bulk Metallic Glasses With Different Length-Scale of Constituent Phases." *Intermetallics* 10, no. 11–12, pp. 1183–90.

Emmerich, H. 2009. "Phase-Field Modelling for Metals and Colloids and Nucleation Therein—An Overview." *Journal of Physics: Condensed Matter* 21, no. 46, p. 464103.

Emmerich, H. 2011. "Phase-Field Models for Colloidal Systems: First Steps, Perspectives and Open Challenges." *Current Opinion in Solid State and Materials Science* 15, no. 3, pp. 83–86.

Emmerich, H., H. Lowen, R. Wittkowski, T. Gruhn, G.I. Toth, G. Tegze, and L. Granasay. 2012. "Phase-Field-Crystal Models for Condensed Matter Dynamics on Atomic Length and Diffusive Time Scales: An Overview." *Advances in Physics* 61, no. 6, pp. 665–743.

Ercolessi, F., M. Parrinello, and E. Tosatti. 1988. "Simulation of Gold in the Glue Model." *Philosophical Magazine A* 58, no. 1, pp. 213–26.

Eshraghi, M., B. Jelinek, and S.D. Felicelli. 2015. "Large-Scale Three-Dimensional Simulation of Dendritic Solidification Using Lattice Boltzmann Method." *JOM* 67, no. 8, pp. 1786–92.

Falk, M.L. 1999. "Molecular-Dynamics Study of Ductile and Brittle Fracture in Model Noncrystalline Solids." *Physical Review B* 60, no. 10, pp. 7062–70.

Fan, C. and A. Inoue. 2000. "Ductility of Bulk Nanocrystalline Composites and Metallic Glasses at Room Temperature." *Applied Physics Letters* 77, no. 1, pp. 46–48.

Fan, C., L.J. Kecskes, D.-C. Qiao, H. Choo, and P.K. Liaw. 2006b. "Properties of As-Cast and Structurally Relaxed Zr-Based Bulk Metallic Glasses." *Journal of Non-Crystalline Solids* 352, no. 2, pp. 174–9.

Fan, C., H.L. Laszlo, J. Kecskes, K. Tao, H. Choo, P.K. Liaw, and C.T. Liu. 2006a. "Mechanical Behavior of Bulk Amorphous Alloys Reinforced by Ductile Particles at Cryogenic Temperatures." *Physical Review Letters* 96, no. 14, p. 145506.

Fan, C., C. Li, and A. Inoue. 2000. "Nanocrystal Composites in Zr–Nb–Cu–Al Metallic Glasses." *Journal of Non-Crystalline Solids* 270, no. 1–3, pp. 28–33.

Fan, C., C. Li, and A. Inoue. 2001. "Effects of Nb Addition on Icosahedral Quasicrystalline Phase Formation and Glass-Forming Ability of Zr–Ni–Cu–Al Metallic Glasses." *Applied Physics Letters* 79, no. 7, pp. 1024–6.

Fan, C., R.T. Ott, and T.C. Hufnagel. 2002. "Metallic Glass Matrix Composite With Precipitated Ductile Reinforcement." *Applied Physics Letters* 81, no. 6, pp. 1020–2.

Fan, Z., T.E. Sparks, F. Liou, A. Jambunathan, Y. Bao, J. Ruan, and J.W. Newkirk. 2007. Numerical simulation of the evolution of solidification microstructure in laser deposition. In *Proceedings of the 18th Annual Solid Freeform Fabrication Symposium*, The University of Texas at Austin, Austin, TX.

Feng, S.D., W. Jiao, S.P. Pan, L. Qi, W. Gao, L.M. Wang, G. Li, M.Z. Ma, and R.P. Liu. 2015. "Transition From Elasticity to Plasticity in Zr35Cu65 Metallic Glasses: A Molecular Dynamics Study." *Journal of Non-Crystalline Solids* 430, pp. 94–98.

Figueroa, I.A., P.A. Carroll, H.A. Davies, H. Jones, and I. Todd. 2007. Preparation of Cu-based bulk metallic glasses by suction casting. In *SP07 Proceedings of 5th Decennial International Conference on Solidification Processing*, ed. H. Jones, pp. 479–82. Shieffield: Department of Engineering Materials, University of Shieffield.

Firstov, G.S., J. Van Humbeeck, and Y.N. Koval. 2004. "High-Temperature Shape Memory Alloys: Some Recent Developments." *Materials Science and Engineering: A* 378, no. 1–2, pp. 2–10.

Flores, K.M. 2006. "Structural Changes and Stress State Effects During Inhomogeneous Flow of Metallic Glasses." *Scripta Materialia* 54, no. 3, pp. 327–32.

Flores, K.M. and R.H. Dauskardt. 2004. "Fracture and Deformation of Bulk Metallic Glasses and Their Composites." *Intermetallics* 12, no. 7–9, pp. 1025–9.

Foiles, S., M. Baskes, and M.S. Daw. 1986. "Embedded-Atom-Method Functions for the FCC Metals Cu, Ag, Au, Ni, Pd, Pt, and Their Alloys." *Physical Review B* 33, no. 12, p. 7983.

Francois, M.M., S.J. Cummins, E.D. Dendy, D.B. Kothe, J.M. Sicilian, and M.W. Williams. 2006. "A Balanced-Force Algorithm for Continuous and Sharp Interfacial Surface Tension Models Within a Volume Tracking Framework." *Journal of Computational Physics* 213, no. 1, pp. 141–73.

Frazier, W.E. 2014. "Metal Additive Manufacturing: A Review." *Journal of Materials Engineering and Performance* 23, no. 6, pp. 1917–28.

Freed, R.L. and J.B. Vander Sande. 1979. "The Effects of Devitrification on the Mechanical Properties of Cu46Zr54 Metallic Glass." *Metallurgical Transactions A* 10, no. 11, pp. 1621–6.

Frijns, A.J.H., S.V. Nedea, A.J. Markvoort, A. A. van Steenhoven, P.A.J. Hilbers. 2004. Molecular dynamics and Monte Carlo simulations for heat transfer in micro and nano-channels. In *Computational Science – ICCS 2004: 4th International Conference, Kraków, Poland, June 6–9, 2004, Proceedings, Part IV*, ed. M. Bubak, et al., pp. 661–6. Berlin, Heidelberg: Springer.

Fu, X.L., Y. Li, and C.A. Schuh. 2007. "Temperature, Strain Rate and Reinforcement Volume Fraction Dependence of Plastic Deformation in Metallic Glass Matrix Composites." *Acta Materialia* 55, no. 9, pp. 3059–71.

Gandin, C.-A., J.-L. Desbiolles, M. Rappaz, and Ph. Thevoz. 1999. "A Three-Dimensional Cellular Automation-Finite Element Model for the Prediction of Solidification Grain Structures." *Metallurgical and Materials Transactions A* 30, no. 12, pp. 3153–65.

Gandin, C.A. and M. Rappaz. 1994. "A Coupled Finite Element-Cellular Automaton Model for the Prediction of Dendritic Grain Structures in Solidification Processes." *Acta Metallurgica et Materialia* 42, no. 7, pp. 2233–46.

Gandin, C.A. and M. Rappaz. 1997. "A 3D Cellular Automaton Algorithm for the Prediction of Dendritic Grain Growth." *Acta Materialia* 45, no. 5, pp. 2187–95.

Gandin, C.A., R.J. Schaefer, and M. Rappax. 1996. "Analytical and Numerical Predictions of Dendritic Grain Envelopes." *Acta Materialia* 44, no. 8, pp. 3339–47.

Ganesan, S. and D.R. Poirier. 1990. "Conservation of Mass and Momentum for the Flow of Interdendritic Liquid During Solidification." *Metallurgical Transactions B* 21, no. 1, pp. 173–81.

Gao, W.-H., X.L. Meng, W. Cai, and L.-C. Zhao. 2015. "Effects of Co and Al Addition on Martensitic Transformation and Microstructure in ZrCu-Based Shape Memory Alloys." *Transactions of Nonferrous Metals Society of China* 25, no. 3, pp. 850–5.

Gargarella, P., S. Pauly, K.K. Song, J. Hu, N.S. Barekar, M.S. Khoshkhoo, A. Teresiak, et al. 2013. "Ti–Cu–Ni Shape Memory Bulk Metallic Glass Composites." *Acta Materialia* 61, no. 1, pp. 151–62.

Georgarakis, K., M. Aljert, Y. Li, A. LeMoulec, F. Charlot, A.R. Yavari, K. Choronokhvostenko, et al. 2008. "Shear Band Melting and Serrated Flow in Metallic Glasses." *Applied Physics Letters* 93, no. 3, p. 031907.

Gibson, I., W.D. Rosen, and B. Stucker. 2010a. "Development of Additive Manufacturing Technology." In *Additive Manufacturing Technologies: Rapid Prototyping to Direct Digital Manufacturing*. Boston, MA: Springer. pp. 36–58.

Gibson, I., W.D. Rosen, and B. Stucker. 2010b. "Medical Applications for Additive Manufacture." In *Additive Manufacturing Technologies: Rapid Prototyping to Direct Digital Manufacturing.*Boston, MA: Springer. pp. 400–14.

Gilbert, A. 2007. *Introduction to Computational Quantum Chemistry: Theory.* University Lecture.

Gilbert, C.J., R.O. Ritchie, and W.L. Johnson. 1997. "Fracture Toughness and Fatigue-Crack Propagation in a Zr–Ti–Ni–Cu–Be Bulk Metallic Glass." *Applied Physics Letters* 71, no. 4, pp. 476–8.

Glardon, R., N. Karapatis, V. Romano, and G.N. Levy. 2001. "Influence of Nd:YAG Parameters on the Selective Laser Sintering of Metallic Powders." *CIRP Annals - Manufacturing Technology* 50, no. 1, pp. 133–6.

Gludovatz, B., S.E. Naleway, R.O. Ritchie, and J.J. Kruzic. 2014. "Size-Dependent Fracture Toughness of Bulk Metallic Glasses." *Acta Materialia* 70, pp. 198–207.

Gong, X., T. Anderson, and K. Chou. 2012. Review on powder-based electron beam additive manufacturing technology. In *ASME/ISCIE 2012 International Symposium on Flexible Automation,* American Society of Mechanical Engineers, New York City.

Gong, X. and K. Chou. 2015. "Phase-Field Modeling of Microstructure Evolution in Electron Beam Additive Manufacturing." *JOM* 67, no. 5, pp. 1176–82.

González, S. 2015. "Role of Minor Additions on Metallic Glasses and Composites." *Journal of Materials Research* 31, no. 1, pp. 76–87.

Gránásy, L. 1993. "Quantitative Analysis of the Classical Nucleation Theory on Glass-Forming Alloys." *Journal of Non-Crystalline Solids* 156, pp. 514–8.

Gránásy, L., L. Ratkai, A. Szallas, B. Korbuly,, G.I. Toth, L. Kornyei, and T. Pusztai. 2014. "Phase-Field Modeling of Polycrystalline Solidification: From Needle Crystals to Spherulites—A Review." *Metallurgical and Materials Transactions A* 45, no. 4, pp. 1694–719.

Greer, A.L. 1993. "Confusion by Design." *Nature* 366, no. 6453, pp. 303–4.

Greer, A.L. 1995. "Metallic Glasses." *Science* 267, no. 5206, pp. 1947–53.

Greer, A.L. 2006. "Liquid Metals: Supercool Order." *Nature Materials* 5, no. 1, pp. 13–14.

Greer, A.L., Y.Q. Cheng, and E. Ma. 2013. "Shear Bands in Metallic Glasses." *Materials Science and Engineering: R: Reports* 74, no. 4, pp. 71–132.

Griffith, M.L., M.T. Ensz, J.D. Puskar, C.V. Robino, J.A. Brooks, J.A. Philliber, J.E. Smugeresky, and W.H. Hofmeiser. 2011. Understanding the microstructure and properties of components fabricated by laser engineered net shaping (LENS). *MRS Proceedings,* p. 625.

Griffith, M., D.M. Keicher, C.L. Atwood, J.A. Romero, J.E. Smugeresky, L.D. Harwell, and D.L. Greene. 1996. Free form fabrication of metallic components using laser engineered net shaping (LENS). In *Proceedings of the Solid Freeform Fabrication Symposium,* University of Texas at Austin Austin, TX.

Gu, X.J., S.J. Poon, and G.J. Shiflet. 2007. "Mechanical Properties of Iron-Based Bulk Metallic Glasses." *Journal of Materials Research* 22, no. 02, pp. 344–51.

Gu, J., M. Song, S. Ni, S. Guo, and Y. He. 2013. "Effects of Annealing on the Hardness and Elastic Modulus of a Cu36Zr48Al8Ag8 Bulk Metallic Glass." *Materials & Design* 47, pp. 706–10.

Guillemot, G., Ch.-A. Gandin, H. Combeau, and R. Heringer. 2004. "A New Cellular Automaton—Finite Element Coupling Scheme for Alloy Solidification." *Modelling and Simulation in Materials Science and Engineering* 12, no. 3, p. 545.

Güntherodt, H.J. 1977. "Metallic Glasses." In *Festkörperprobleme 17: Plenary Lectures of the Divisions "Semiconductor Physics" "Metal Physics" "Low Temperature Physics" "Thermodynamics and Statistical Physics" "Crystallography" "Magnetism" "Surface Physics" of the German Physical Society Münster, March 7–12, 1977*, ed. Treusch. Berlin, Heidelberg: Springer, pp. 25–53.

Guo, G.-Q., S.-Y. Wu, S. Luo, and L. Yang. 2015a. "Detecting Structural Features in Metallic Glass via Synchrotron Radiation Experiments Combined with Simulations." *Metals* 5, no. 4, p. 2093.

Guo, G.-Q., S.-Y. Wu, S. Luo, and L. Yang. 2015b. "How Can Synchrotron Radiation Techniques Be Applied for Detecting Microstructures in Amorphous Alloys?" *Metals* 5, no. 4, p. 2048.

Guo, G.-Q., S.-Y. Wu, and L. Yang. 2016. "Structural Origin of the Enhanced Glass-Forming Ability Induced by Microalloying Y in the ZrCuAl Alloy." *Metals* 6, no. 4, p. 67.

Guo, G.-Q., L. Yang, S.Y. Wu, Q.S. Zeng, C.J. Sun, and Y.G. Wang. 2016. "Structure-Induced Microalloying Effect in Multicomponent Alloys." *Materials & Design* 103, pp. 308–14.

Guo, H., P.F. Yan, Y.B. Wang, J. Tan, Z.F. Zhang, M.L. Sui, and E. Ma. 2007. "Tensile Ductility and Necking of Metallic Glass." *Nature Materials* 6, no. 10, pp. 735–9.

Guo, S.F., J.L. Qiu, P. Yu, S.H. Xie, and W. Chen. 2014. "Fe-Based Bulk Metallic Glasses: Brittle or Ductile?" *Applied Physics Letters* 105, no. 16, p. 161901.

Gusarov, A.V. and I. Smurov. 2010. "Modeling the Interaction of Laser Radiation With Powder Bed at Selective Laser Melting." *Physics Procedia* 5, pp. 381–94.

Hao, S., L. Cui, D. Jiang, X. Han, Y. Ren, J. Jiang, Y. Liu, et al. 2013. "A Transforming Metal Nanocomposite With Large Elastic Strain, Low Modulus, and High Strength." *Science* 339, no. 6124, pp. 1191–4.

Hao, T., Z. Zhou, Y. Nie, L. Zhu, Y. Wei, and S. Li. 2016. "Molecular Simulations of Crystallization Behaviors of Polymers Grafted on Two-Dimensional Filler." *Polymer* 100, pp. 10–18.

Harooni, A., A.M. Nasiri, A.P. Gerlich, A. Khajepour, A. Khalifa, and J.M. King. 2016. "Processing Window Development for Laser Cladding of Zirconium on Zirconium Alloy." *Journal of Materials Processing Technology* 230, pp. 263–71.

Hays, C.C., C.P. Kim, and W.L. Johnson. 2000. "Microstructure Controlled Shear Band Pattern Formation and Enhanced Plasticity of Bulk Metallic Glasses

Containing *In Situ* Formed Ductile Phase Dendrite Dispersions." *Physical Review Letters* 84, no. 13, pp. 2901–4.

Hays, C.C., C.P. Kim, and W.L. Johnson. 2001. "Improved Mechanical Behavior of Bulk Metallic Glasses Containing In Situ Formed Ductile Phase Dendrite Dispersions." *Materials Science and Engineering: A* 304–306, pp. 650–5.

He, G., J. Eckert, and W. Löser. 2003. "Stability, Phase Transformation and Deformation Behavior of Ti-Base Metallic Glass and Composites." *Acta Materialia* 51, no. 6, pp. 1621–31.

He, G., J. Eckert, W. Loser, and L. Schultz. 2003. "Novel Ti-Base Nanostructure-Dendrite Composite With Enhanced Plasticity." *Nature Materials* 2, no. 1, pp. 33–37.

He, Y., R.B. Schwarz, and J.I. Archuleta. 1996. "Bulk Glass Formation in the Pd–Ni–P System." *Applied Physics Letters* 69, no. 13, pp. 1861–3.

Hebert, R.J. 2016. "Viewpoint: Metallurgical Aspects of Powder Bed Metal Additive Manufacturing." *Journal of Materials Science* 51, no. 3, pp. 1165–75.

Hebi, Y. and D.F. Sergio. 2009. "A Cellular Automaton Model for Dendrite Growth in Magnesium Alloy AZ91." *Modelling and Simulation in Materials Science and Engineering* 17, no. 7, p. 075011.

Hofmann, D.C. 2010. "Shape Memory Bulk Metallic Glass Composites." Science 329, no. 5997, pp. 1294–5.

Hofmann, D.C. and W.L. Johnson. 2010. "Improving Ductility in Nanostructured Materials and Metallic Glasses:'Three Laws'." In *Materials Science Forum,* Trans Tech Publications, Switzerland.

Hofmann, D.C., J. Kolodziejska, S. Roberts, R. Otis, R.P. Dilon, J.-O. Suh, Z.-K. Liu, and J.P. Borgonia. 2014a. "Compositionally Graded Metals: A New Frontier of Additive Manufacturing." *Journal of Materials Research* 29, no. 17, pp. 1899–910.

Hofmann, D.C., H. Kozachkov, H.E. Khalifa, J.P. Schramm, M.D. Demetriou, K.S. Vecchio, and W.L. Johnson, 2009. "Semi-Solid Induction Forging of Metallic Glass Matrix Composites." *JOM* 61, no. 12, pp. 11–17.

Hofmann, D.C. and S.N. Roberts. 2015. "Microgravity Metal Processing: From Undercooled Liquids to Bulk Metallic Glasses." *npj Microgravity* 1, p. 15003.

Hofmann, D.C., S. Roberts, R. Otis, J. Kolodziejska, R.P. Dillion, J.O. Suh, A.A. Shapiro, Z.-K. Liu, and J.P. Borgonia. 2014b. "Developing Gradient Metal Alloys through Radial Deposition Additive Manufacturing." *Scientific Reports* 4, p. 5357.

Hofmann, D.C., J.Y. Suh, A. Wiest, G. Duan, M.L. Lind, M.D. Demetriou, and W.L. Johnson. 2008a. "Designing Metallic Glass Matrix Composites With High Toughness and Tensile Ductility." *Nature* 451, no. 7182, pp. 1085–9.

Hofmann, D.C., J.Y. Suh, A. Wiest, and W. Johnson. 2008b. "New Processing Possibilities for Highly Toughened Metallic Glass Matrix Composites With Tensile Ductility." *Scripta Materialia* 59, no. 7, pp. 684–7.

Hou, B., Y.L. Li, L.Q. Xing, C.-S. Chen, H.C. Kou, and J.S. Li. 2007. "Dynamic and Quasi-Static Mechanical Properties of Fibre-Reinforced Metallic Glass at Different Temperatures." *Philosophical Magazine Letters* 87, no. 8, pp. 595–601.

Howell, J.R. 1998. "The Monte Carlo Method in Radiative Heat Transfer." *Journal of Heat Transfer* 120, no. 3, pp. 547–60.

Hu, X., S.C. Ng, Y.P. Feng, and Y. Li. 2003. "Glass Forming Ability and In-Situ Composite Formation in Pd-Based Bulk Metallic Glasses." *Acta Materialia* 51, no. 2, pp. 561–72.

Hu, Y., K.C. Chan, L. Liu, and Y.Z. Yang. 2010. "Monte Carlo Simulation of Dual Magnetic Phase Behavior in Bulk Metallic Glasses." *Journal of Magnetism and Magnetic Materials* 322, no. 17, pp. 2567–70.

Huang, J.C., J.P. Chu, and J.S.C. Jang. 2009. "Recent Progress in Metallic Glasses in Taiwan." *Intermetallics* 17, no. 12, pp. 973–87.

Huang, L., C. Zhu, C.I. Muntele, T. Zhang, P.K. Liaw, and W. He. 2015. "Surface Engineering of a Zr-Based Bulk Metallic Glass with Low Energy Ar- or Ca-Ion Implantation." *Materials Science and Engineering: C* 47, pp. 248–55.

Huang, R., Z. Suo, J.H. Prevost, and W.D. Nix. 2002. "Inhomogeneous Deformation in Metallic Glasses." *Journal of the Mechanics and Physics of Solids* 50, no. 5, pp. 1011–27.

Huang, Y.J., J. Shen, and J.F. Sun. 2007. "Bulk Metallic Glasses: Smaller is Softer." *Applied Physics Letters* 90, no. 8, p. 081919.

Hufnagel, T.C. 2006. "Preface to the Viewpoint Set on Mechanical Behavior of Metallic Glasses." *Scripta Materialia* 54, no. 3, pp. 317–9.

Hufnagel, T.C., C. Fan, R.T. Ott, J. Li, and B. Brennan. 2002a. "Controlling Shear Band Behavior in Metallic Glasses Through Microstructural Design." *Intermetallics* 10, no. 11–12, pp. 1163–6.

Hufnagel, T.C., T. Jiao, Y. Li, L.-Q. Xing, and K.T. Ramesh. 2002b. "Deformation and Failure of Zr57Ti5Cu20Ni8Al10 Bulk Metallic Glass Under Quasi-Static and Dynamic Compression." *Journal of Materials Research* 17, no. 06, pp. 1441–5.

Hufnagel, T.C., R.T. Ott, and J. Almer. 2006. "Structural Aspects of Elastic Deformation of a Metallic Glass." *Physical Review B* 73, no. 6, p. 064204.

Hufnagel, T.C., C.A. Schuh, and M.L. Falk. 2016. "Deformation of Metallic Glasses: Recent Developments in Theory, Simulations, and Experiments." *Acta Materialia* 109, pp. 375–93.

Hufnagel, T.C., U.K. Vempati, and J.D. Almer. 2013. "Crack-Tip Strain Field Mapping and the Toughness of Metallic Glasses." *PLoS ONE* 8, no. 12, p. e83289.

Hui, X., J. Yu, M. Wang, W. Dong, and G. Chen. 2006. "Wetting Angle and Infiltration Velocity of Zr Base Bulk Metallic Glass Composite." *Intermetallics* 14, no. 8–9, pp. 931–5.

Hui, Z., H. Yizhu, Y. Xiaomin, and P. Ye. 2010. "Microstructure and Age Characterization of Cu15Ni–8Sn Alloy Coatings by Laser Cladding." *Applied Surface Science* 256, no. 20, pp. 5837–42.

Hunt, J.D. 1984. "Steady State Columnar and Equiaxed Growth of Dendrites and Eutectic." *Materials Science and Engineering* 65, no. 1, pp. 75–83.

Hwang, J., Z.H. Melgarejo, Y.E. Kalay, I. Kalay, M.J. Kramer, D.S. Stone, and P.M. Voyles. 2012. "Nanoscale Structure and Structural Relaxation in $Zr_{50}Cu_{45}Al_5$ Bulk Metallic Glass." *Physical Review Letters* 108, no. 19, p. 195505.

Inoue, A. 1995. "High Strength Bulk Amorphous Alloys with Low Critical Cooling Rates (*Overview*)." *Materials Transactions, JIM* 36, no. 7, pp. 866–75.

Inoue, A. 2000. "Stabilization of Metallic Supercooled Liquid and Bulk Amorphous Alloys." *Acta Materialia* 48, no. 1, pp. 279–306.

Inoue, A. 2001. "Bulk Amorphous and Nanocrystalline Alloys With High Functional Properties." *Materials Science and Engineering: A* 304–306, pp. 1–10.

Inoue, A. 2015. "Bulk Glassy Alloys: Historical Development and Current Research." *Engineering* 1, no. 2, pp. 185–91.

Inoue, A., F.L. Kong, S.L. Zhu, E. Shalaan, and F.M. Al-Marzouki. 2015. "Production Methods and Properties of Engineering Glassy Alloys and Composites." *Intermetallics* 58, pp. 20–30.

Inoue, A., N. Nishiyama, and T. Matsuda. 1996. "Preparation of Bulk Glassy $Pd_{40}Ni_{10}Cu_{30}P_{20}$ Alloy of 40 mm in Diameter by Water Quenching." *Materials Transactions, JIM* 37, no. 2, pp. 181–4.

Inoue, A., B. Shen, and A. Takeuchi. 2006. "Developments and Applications of Bulk Glassy Alloys in Late Transition Metal Base System." *Materials Transactions* 47, no. 5, pp. 1275–85.

Inoue, A. and A. Takeuchi. 2002. "Bulk Amorphous, Nano–Crystalline and Nano–Quasicrystalline Alloys IV. Recent Progress in Bulk Glassy Alloys." *Materials Transactions* 43, no. 8, pp. 1892–906.

Inoue, A. and A. Takeuchi. 2011. "Recent Development and Application Products of Bulk Glassy Alloys." *Acta Materialia* 59, no. 6, pp. 2243–67.

Inoue, A. and T. Zhang. 1995. "Fabrication of Bulky Zr-Based Glassy Alloys by Suction Casting into Copper Mold." *Materials Transactions, JIM* 36, no. 9, pp. 1184–7.

Inoue, A. and T. Zhang. 1996. "Fabrication of Bulk Glassy $Zr_{55}Al_{10}Ni_5Cu_{30}$ Alloy of 30 mm in Diameter by a Suction Casting Method." *Materials Transactions, JIM* 37, no. 2, pp. 185–7.

Inoue, A., T. Zhang, and E. Makabe. 1998. "Production Methods of Metallic Glasses by a Suction Casting Method." *Google Patents*.

Inoue, A., T. Zhang, and T. Masumoto. 1990. "Zr Al–Ni Amorphous Alloys With High Glass Transition Temperature and Significant Supercooled Liquid Region." *Materials Transactions, JIM* 31, no. 3, pp. 177–83.

Inoue, A., T. Zhang, and T. Masumoto. 1993. "Glass-Forming Ability of Alloys." *Journal of Non-Crystalline Solids* 156, pp. 473–80.

James, P.F. 1975. "Liquid-Phase Separation in Glass-Forming Systems." *Journal of Materials Science* 10, no. 10, pp. 1802–25.

Jang, D. and J.R. Greer. 2010. "Transition From a Strong-Yet-Brittle to a Stronger-and-Ductile State by Size Reduction of Metallic Glasses." *Nature Materials* 9, no. 3, pp. 215–9.

Janot, C. 1993. "The Structure of Quasicrystals." *Journal of Non-Crystalline Solids* 156, pp. 852–64.

Javid, F.A., N. Mattern, S. Pauly, and J. Eckert. 2011. "Martensitic Transformation and Thermal Cycling Effect in Cu–Co–Zr Alloys." *Journal of Alloys and Compounds* 509, Supplement 1, pp. S334–7.

Jeantette, F.P., D.M. Keicher, J.A. Romero, and L.P. Schanwald. 2000. "Method and System for Producing Complex-Shape Objects." *Google Patents*.

Jeon, C., D.J. Ha, C.P. Kim, and S. Lee. 2012. "Effects of Dendrite Size on Tensile Deformation Behavior in Zr-Based Amorphous Matrix Composites Containing Ductile Dendrites." *Metallurgical and Materials Transactions A* 43, no. 10, pp. 3663–74.

Jeon, C., H. Lee, C.P. Kim, S.H. Joo, S.H. Kim, and S. Lee. 2015. "Effects of Effective Dendrite Size on Tensile Deformation Behavior in Ti-Based Dendrite-Containing Amorphous Matrix Composites Modified from Ti–6Al–4V Alloy." *Metallurgical and Materials Transactions A* 46, no. 1, pp. 235–50.

Jeong, H.G., S.J. Yoo, and W.J. Kim. 2009. "Micro-Forming of Zr65Al10Ni10Cu15 Metallic Glasses Under Superplastic Condition." *Journal of Alloys and Compounds* 483, no. 1–2, pp. 283–5.

Jia, P., H. Guo, Y. Li, J. Xu, and E. Ma. 2006a. "A New Cu–Hf–Al Ternary Bulk Metallic Glass with High Glass Forming Ability and Ductility." *Scripta Materialia* 54, no. 12, pp. 2165–8.

Jia, W., Z. Peng, Z. Wang, X. Ni, and C.Y. Wang. 2006b. "The Effect of Femtosecond Laser Micromachining on the Surface Characteristics and Subsurface Microstructure of Amorphous FeCuNbSiB Alloy." *Applied Surface Science* 253, no. 3, pp. 1299–303.

Jia, W., D. Zhang, X. Li, L. Chai, Z. Zang, and C. Wang. 2008. "Heat Effects of Amorphous FeCuNbSiB Alloy Ablated with Femtosecond Laser." Thin Solid Films 516, no. 8, pp. 2260–3.

Jiang, F., D.H. Zhang, L.C. Zhang, Z.B. Zhnag, L. He, J. Sun, and Z.F. Zhnag. 2007. "Microstructure Evolution and Mechanical Properties of Cu46Zr47Al7 Bulk Metallic Glass Composite Containing CuZr Crystallizing Phases." *Materials Science and Engineering: A* 467, no. 1–2, pp. 139–45.

Jiang, J.-Z., D. Hofmann, D.J. Jarvis, and H.-J. Fecht. 2015. "Low-Density High-Strength Bulk Metallic Glasses and Their Composites: A Review." *Advanced Engineering Materials* 17, no. 6, pp. 761–80.

Jiang, M.Q. and L.H. Dai. 2010. "Short-Range-Order Effects on Intrinsic Plasticity of Metallic Glasses." *Philosophical Magazine Letters* 90, no. 4, pp. 269–77.

Jiang, W.H. and M. Atzmon. 2006. "Mechanically-Assisted Nanocrystallization and Defects in Amorphous Alloys: A High-Resolution Transmission Electron Microscopy Study." *Scripta Materialia* 54, no. 3, pp. 333–6.

Jiang, W.H., G.J. Fan, H. Choo, and P.K. Liaw. 2006. "Ductility of a Zr-Based Bulk-Metallic Glass With Different Specimen's Geometries." *Materials Letters* 60, no. 29–30, pp. 3537–40.

Jiang, W.H., G.J. Fan, F.X. Liu, G.Y. Wang, H. Choo, and P.K. Liaw. 2008. "Spatiotemporally Inhomogeneous Plastic Flow of a Bulk-Metallic Glass." *International Journal of Plasticity* 24, no. 1, pp. 1–16.

Jiang, W.H., F.E. Pinkerton, and M. Atzmon. 2005. "Deformation-Induced Nanocrystallization: A Comparison of Two Amorphous Al-Based Alloys." *Journal of Materials Research* 20, no. 03, pp. 696–702.

Jiang, Y. and K. Qiu. 2015. "Computational Micromechanics Analysis of Toughening Mechanisms of Particle-Reinforced Bulk Metallic Glass Composites." *Materials & Design (1980–2015)* 65, pp. 410–6.

Johnson, W.L. 1999. "Bulk Glass-Forming Metallic Alloys: Science and Technology." *MRS Bulletin* 24, no. 10, pp. 42–56.

Joseph, J., T. Jarvis, X. Wu, N. Stanford, P. Hodgson, and D.M. Fabijanic. 2015. "Comparative Study of the Microstructures and Mechanical Properties of Direct Laser Fabricated and Arc-Melted AlxCoCrFeNi High Entropy Alloys." *Materials Science and Engineering: A* 633, pp. 184–93.

Jung, H.Y., S.J. Choi, K.G. Prashanth, M. Stoica, S. Scudino, S. Yi, U. Kuhn, D.H. Kim, K.B. Kim, and J. Eckert. 2015b. "Fabrication of Fe-Based Bulk Metallic Glass by Selective Laser Melting: A Parameter Study." *Materials & Design* 86, pp. 703–8.

Jung, H.Y., M. Stoica, S. Yi, D.H. Kim, and J. Eckert. 2015a. "Crystallization Kinetics of Fe76.5−x C6.0Si3.3B5.5P8.7Cu x (x = 0, 0.5, and 1 at. pct) Bulk Amorphous Alloy." *Metallurgical and Materials Transactions A* 46, no. 6, pp. 2415–21.

Kahnert, M., S. Lutzmann, and M. Zaeh. 2007. Layer formations in electron beam sintering. In *Solid Freeform Fabrication Symposium,* University of Texas, Austin, TX.

Kandis, M. and T.L. Bergman. 1999. "A Simulation Based Correlation of the Density and Thermal Conductivity of Objects Produced by Laser Sintering of Polymer Powders." *Journal of Manufacturing Science and Engineering* 122, no. 3, pp. 439–44.

Kathuria, Y.P. 1999. "Microstructuring by Selective Laser Sintering of Metallic Powder." *Surface and Coatings Technology* 116–119, pp. 643–7.

Kawahito, Y., T. Terajima, H. Kimura, T. Kuroda, K. Nakata, S. Katayama, and A. Inoue. 2008. "High-Power Fiber Laser Welding and Its Application to Metallic Glass Zr55Al10Ni5Cu30." *Materials Science and Engineering: B* 148, no. 1–3, pp. 105–9.

Kelly, P. and M.-X. Zhang. 1999. "Edge-to-Edge Matching – A New Approach to the Morphology and Crystallography of Precipitates." In *Materials Forum,* Trans Tech Publications, Switzerland.

Kelly, P.M. and M.-X. Zhang. 2006. "Edge-to-Edge Matching — The Fundamentals." *Metallurgical and Materials Transactions A* 37, no. 3, pp. 833–9.

Kelton, K.F. 1998. "A New Model for Nucleation in Bulk Metallic Glasses." *Philosophical Magazine Letters* 77, no. 6, pp. 337–44.

Kempen, K., L. Thijis, J.V. Humbeeck, and J.-P. Kruth. 2012. "Mechanical Properties of AlSi10Mg Produced by Selective Laser Melting." *Physics Procedia* 39, pp. 439–46.

Khademian, N. and R. Gholamipour. 2010. "Fabrication and Mechanical Properties of a Tungsten Wire Reinforced Cu–Zr–Al Bulk Metallic Glass Composite." *Materials Science and Engineering: A* 527, no. 13–14, pp. 3079–84.

Khairallah, S.A., A.T. Anderson, A. Rubenchik, and W.A. King. 2016. "Laser Powder-Bed Fusion Additive Manufacturing: Physics of Complex Melt Flow and Formation Mechanisms of Pores, Spatter, and Denudation Zones." *Acta Materialia* 108, pp. 36–45.

Khalifa, H.E. 2009. *Bulk Metallic Glasses and Their Composites: Composition Optimization, Thermal Stability, and Microstructural Tunability* [PhD Thesis]. San Diego, CA: University of California.

Kim, D.H., W.T. Kim, E.S. Park, N. Mattern, and J. Eckert. 2013. "Phase Separation in Metallic Glasses." *Progress in Materials Science* 58, no. 8, pp. 1103–72.

Kim, J., D. Lee, S. Shin, and C. Lee. 2006. "Phase Evolution in Cu54Ni6Zr22Ti18 Bulk Metallic Glass Nd:YAG Laser Weld." *Materials Science and Engineering: A* 434, no. 1–2, pp. 194–201.

Kim, C.P., Y.S. Oh, S. Lee, and N.J. Kim. 2011. "Realization of High Tensile Ductility in a Bulk Metallic Glass Composite by the Utilization of Deformation-Induced Martensitic Transformation." *Scripta Materialia* 65, no. 4, pp. 304–7.

Kim, J.H., C. Lee, D.M. Lee, J.H. Sun, S.Y. Shin, and J.C. Bae. 2007. "Pulsed Nd:YAG Laser Welding of Cu54Ni6Zr22Ti18 Bulk Metallic Glass." *Materials Science and Engineering: A* 449–451, pp. 872–5.

Kim, J.-H., J.S. Park, E.S. Park, W.T. Kim, and D.H. Kim. 2005a. "Estimation of Critical Cooling Rates for Glass Formation in Bulk Metallic Glasses Through Non-Isothermal Thermal Analysis." *Metals and Materials International* 11, no. 1, pp. 1–9.

Kim, Y.C., D.H. Bae, W.T. Kim, and D.H. Kim. 2003. "Glass Forming Ability and Crystallization Behavior of Ti-Based Amorphous Alloys with High Specific Strength." *Journal of Non-Crystalline Solids* 325, no. 1–3, pp. 242–50.

Kim, K.B., J. Das, F. Baier, and J. Eckert. 2005b. "Heterogeneous Distribution of Shear Strains in Deformed Ti66.1Cu8Ni4.8Sn7.2Nb13.9 Nanostructure-Dendrite Composite." *Physica Status Solidi (a)* 202, no. 13, pp. 2405–12.

Kimura, H. and T. Masumoto. 1975. "Fracture Toughness of Amorphous Metals." *Scripta Metallurgica* 9, no. 3, pp. 211–21.

King, W., A.T. Anderson, R.M. Ferencz, N.E. Hodge, C. Kamath, and S.A. Khairallah. 2015. "Overview of Modelling and Simulation of Metal Powder Bed Fusion Process at Lawrence Livermore National Laboratory." *Materials Science and Technology* 31, no. 8, pp. 957–68.

Kingery, W.D. 1960. *Introduction to Ceramics*. New York, NY: John Wiley and Sons.

Klement, W., Willens, R.H., and P.O.L. Duwez. 1960. "Non-Crystalline Structure in Solidified Gold-Silicon Alloys." *Nature* 187, no. 4740, pp. 869–70.

Koptyug, A., L.-E. Rännar, and M. Bäckström. 2013. Bulk metallic glass manufacturing using electron beam melting. In *International Conference on Additive Manufacturing & 3D Printing*, Nottingham, UK, July 2013.

Körner, C. 2016. "Additive Manufacturing of Metallic Components by Selective Electron Beam Melting — A Review." *International Materials Reviews* 61, no. 5, pp. 361–77.

Krämer, L., K.S. Kormout, D. Setman, Y. Champion, and R. Pippan. 2015. "Production of Bulk Metallic Glasses by Severe Plastic Deformation." *Metals* 5, no. 2, p. 720.

Kremeyer, K. 1998. "Cellular Automata Investigations of Binary Solidification." *Journal of Computational Physics* 142, no. 1, pp. 243–63.

Krivtsov, A.M. and M. Wiercigroch. 2001. "Molecular Dynamics Simulation of Mechanical Properties for Polycrystal Materials." *Materials Physics and Mechanics* 3, pp. 45–51.

Kroese, D.P., T. Brereton,, T. Taimre, and Z.I. Botev. 2014. "Why the Monte Carlo Method Is so Important Today." *Wiley Interdisciplinary Reviews: Computational Statistics* 6, no. 6, pp. 386–92.

Krone, S.M. 2004. "Spatial Models: Stochastic and Deterministic." *Mathematical and Computer Modelling* 40, no. 3, pp. 393–409.

Kruth, J.P., L. Froyen, J.V. Vaerenberg, P. Mercelis, M. Rombouts, and B. Lauwers. 2004. "Selective Laser Melting of Iron-Based Powder." *Journal of Materials Processing Technology* 149, no. 1–3, pp. 616–22.

Kühn, U., J. Eckert, N. Mattern, and L. Schultz. 2002. "ZrNbCuNiAl Bulk Metallic Glass Matrix Composites Containing Dendritic BCC Phase Precipitates." *Applied Physics Letters* 80, no. 14, pp. 2478–80.

Kühn, U., J. Eckert, N. Mattern, and L. Schultz. 2004. "Microstructure and Mechanical Properties of Slowly Cooled Zr–Nb–Cu–Ni–Al Composites With Ductile BCC Phase." *Materials Science and Engineering: A* 375–377, pp. 322–6.

Kui, H.W., A.L. Greer, and D. Turnbull. 1984. "Formation of Bulk Metallic Glass by Fluxing." *Applied Physics Letters* 45, no. 6, pp. 615 6.

Kumar, S. 2003. "Selective Laser Sintering: A Qualitative and Objective Approach." *JOM* 55, no. 10, pp. 43–47.

Kumar, G., A. Desai, and J. Schroers. 2011. "Bulk Metallic Glass: The Smaller the Better." *Advanced Materials* 23, no. 4, pp. 461–76.

Kündig, A.A., M. Ohnuma, D.H. Ping, T. Ohkubo, and K. Hono. 2004. "In Situ Formed Two-Phase Metallic Glass With Surface Fractal Microstructure." *Acta Materialia* 52, no. 8, pp. 2441–8.

Kurz, W. and D.J. Fisher. 1986. *Fundamentals of Solidification.* Switzerland: Trans Tech Publications.

Kurz, W., B. Giovanola, and R. Trivedi. 1986. "Theory of Microstructural Development During Rapid Solidification." *Acta Metallurgica* 34, no. 5, pp. 823–30.

Lad, K.N. 2014. "Correlation Between Atomic-Level Structure, Packing Efficiency and Glass-Forming Ability in Cu–Zr Metallic Glasses." *Journal of Non-Crystalline Solids* 404, pp. 55–60.

Lan, S., M. Blodgett, K.F. Kelton, J.L. Ma, J. Fan, and X.L. Wang. 2016. "Structural Crossover in a Supercooled Metallic Liquid and the Link to a Liquid-to-Liquid Phase Transition." *Applied Physics Letters* 108, no. 21, p. 211907.

Lee, Y. 2015. *Simulation of Laser Additive Manufacturing and its Applications.* The Ohio State University.

Lee, J.-C., Y.C. Kim, J.P. Ahn, S. Lee, and B.J. Lee. 2004. "Strain Hardening of an Amorphous Matrix Composite Due to Deformation-Induced Nanocrystallization During Quasistatic Compression." *Applied Physics Letters* 84, no. 15, pp. 2781–3.

Lee, M.L., Y. Li, and C.A. Schuh. 2004. "Effect of a Controlled Volume Fraction of Dendritic Phases on Tensile and Compressive Ductility in La-Based Metallic Glass Matrix Composites." *Acta Materialia* 52, no. 14, pp. 4121–31.

Lee, K., C.Y. Son, S.-B. Lee, S.K. Lee, and S. Lee. 2010. "Direct Observation of Microfracture Process in Metallic-Continuous-Fiber-Reinforced Amorphous Matrix Composites Fabricated by Liquid Pressing Process." *Materials Science and Engineering: A* 527, no. 4–5, pp. 941–6.

Lee, D.M., J.H. Sun, D.H. Kang, S.Y. Shin, G. Welsch, and C.H. Lee. 2012. "A Deep Eutectic Point in Quaternary Zr–Ti–Ni–Cu System and Bulk Metallic Glass Formation Near the Eutectic Point." *Intermetallics* 21, no. 1, pp. 67–74.

Lee, Y.S. and W. Zhang. 2016. "Modeling of Heat Transfer, Fluid Flow and Solidification Microstructure of Nickel-Base Superalloy Fabricated by Laser Powder Bed Fusion." *Additive Manufacturing* 12, pp. 178–88.

Leng, Y. and T.H. Courtney. 1991. "Multiple Shear Band Formation in Metallic Glasses in Composites." *Journal of Materials Science* 26, no. 3, pp. 588–92.

Leuders, S., M. Thone, A. Riemer, T. Niendorf, T. Troster, H.A. Richard, and H.J. Maier. 2013. "On the Mechanical Behaviour of Titanium Alloy TiAl6V4 Manufactured by Selective Laser Melting: Fatigue Resistance and Crack Growth Performance." *International Journal of Fatigue* 48, pp. 300–7.

Levine, D. and P.J. Steinhardt. 1985. "Proceedings of the International Conference on the Theory of the Structures of Non-Crystalline Solids Quasicrystals." *Journal of Non-Crystalline Solids* 75, no. 1, pp. 85–89.

Li, N., W. Chen, and L. Liu. 2016. "Thermoplastic Micro-Forming of Bulk Metallic Glasses: A Review." *JOM* 68, no. 4, pp. 1246–61.

Li, B., Z.Y. Li, J.G. Xiong, L. Xing, D. Wang, and Y. Li. 2006. "Laser Welding of Zr45Cu48Al7 Bulk Glassy Alloy." *Journal of Alloys and Compounds* 413, no. 1–2, pp. 118–21.

Li, F., X.J. Liu, H.Y. Hou, G. Chen, and G.L. Chen. 2011. "Structural Origin Underlying Poor Glass Forming Ability of Al Metallic Glass." *Journal of Applied Physics* 110, no. 1, p. 013519.

Li, P., G. Wang, D. Ding, and J. Shen. 2012. "Glass Forming Ability and Thermodynamics of New Ti–Cu–Ni–Zr Bulk Metallic Glasses." *Journal of Non-Crystalline Solids* 358, no. 23, pp. 3200–4.

Li, P., G. Wang, D. Ding, and J. Shen. 2014a. "Glass Forming Ability, Thermodynamics and Mechanical Properties of Novel Ti–Cu–Ni–Zr–Hf Bulk Metallic Glasses." *Materials & Design* 53, pp. 145–51.

Li, H.Q., J.H. Yan, and H.J. Wu. 2009. "Modelling and Simulation of Bulk Metallic Glass Production Process With Suction Casting." *Materials Science and Technology* 25, no. 3, pp. 425–31.

Li, X.P., C.W. Kang, H. Huang, L.C. Zhang, and T.B. Sercombe. 2014b. "Selective Laser Melting of an Al86Ni6Y4.5Co2La1.5 Metallic Glass: Processing, Microstructure Evolution and Mechanical Properties." *Materials Science and Engineering: A* 606, pp. 370–9.

Li, X.P., C.W. Kang, H. Huang, and T. Sercombe. 2014c. "The Role of a Low-Energy–Density Re-Scan in Fabricating Crack-Free Al85Ni5Y6Co2Fe2

Bulk Metallic Glass Composites Via Selective Laser Melting." *Materials & Design* 63, pp. 407–11.

Li, X.P., M. Roberts, Y.J. Liu, C.W. Kang, H. Huang, and T. Sercombe. 2015. "Effect of Substrate Temperature on the Interface Bond Between Support and Substrate During Selective Laser Melting of Al–Ni–Y–Co–La Metallic Glass." *Materials & Design (1980–2015)* 65, pp. 1–6.

Li, X.P., M.P. Roberts, S. O'Keeffe, and T. Sercobe. 2016. "Selective Laser Melting of Zr-Based Bulk Metallic Glasses: Processing, Microstructure and Mechanical Properties." *Materials & Design* 112, pp. 217–26.

Li, Y. and D. Gu. 2014. "Thermal Behavior During Selective Laser Melting of Commercially Pure Titanium Powder: Numerical Simulation and Experimental Study." *Additive Manufacturing* 1–4, pp. 99–109.

Li, Y., S.C. Ng, C.K. Ong, H.H. Hng,, and T.T. Goh. 1997. "Glass Forming Ability of Bulk Glass Forming Alloys." *Scripta Materialia* 36, no. 7, pp. 783–7.

Lin, H.-K., C.-J. Lee, T.-T. Hu, C.-H. Li, and J.C. Huang. 2012. "Pulsed Laser Micromachining of Mg–Cu–Gd Bulk Metallic Glass." *Optics and Lasers in Engineering* 50, no. 6, pp. 883–6.

Lindgren, L.-E., A. Lundback, M. Fisk, R. Pederson, and J. Andersson. 2016. "Simulation of Additive Manufacturing Using Coupled Constitutive and Microstructure Models." *Additive Manufacturing* 12, pp. 144–58.

Liu, C., E. Pineda, and D. Crespo. 2015. "Mechanical Relaxation of Metallic Glasses: An Overview of Experimental Data and Theoretical Models." *Metals* 5, no. 2, pp. 1073.

Liu, J., X. Yuan, H. Zhang, H. Fu, and Z. Hu. 2010a. "Microstructure and Compressive Properties of In-Situ Martensite CuZr Phase Reinforced ZrCu-NiAl Metallic Glass Matrix Composite." *Materials Transactions* 51, no. 5, pp. 1033–7.

Liu, L.F., L.H. Dai, Y.L. Bai, B.C. Wei, and J. Eckert. 2005. "Behavior of Multiple Shear Bands in Zr-Based Bulk Metallic Glass." *Materials Chemistry and Physics* 93, no. 1, pp. 174–7.

Liu, T., P. Shen, F. Qiu, T. Zhang, and Q. Jiang. 2009. "Microstructures and Mechanical Properties of ZrC Reinforced (Zr–Ti)–Al–Ni–Cu Glassy Composites by an In Situ Reaction." *Advanced Engineering Materials* 11, no. 5, pp. 392–8.

Liu, X.J., Y. Xu, X. Hui, Z.P. Lu, F. Li, G.L. Chen, J. Lu, and C.T. Liu. 2010b. "Metallic Liquids and Glasses: Atomic Order and Global Packing." *Physical Review Letters* 105, no. 15, p. 155501.

Liu, Z., R. Li, G. Liu, K. Song, S. Pauly, T. Zhang, and J. Eckert. 2012a. "Pronounced Ductility in CuZrAl Ternary Bulk Metallic Glass Composites With Optimized Microstructure Through Melt Adjustment." *AIP Advances* 2, no. 3, p. 032176.

Liu, Z., R. Li, G. Liu, W. Su, H. Wang, Y. Li, M. Shi, X. Luo, G. Wu, and T. Zhang. 2012b. "Microstructural Tailoring and Improvement of Mechanical Properties in CuZr-Based Bulk Metallic Glass Composites." *Acta Materialia* 60, no. 6–7, pp. 3128–39.

Liu, Z. and H. Qi. 2014. "Numerical Simulation of Transport Phenomena for a Double-Layer Laser Powder Deposition of Single-Crystal Superalloy." *Metallurgical and Materials Transactions A* 45, no. 4, pp. 1903–15.

Liu, Z.Q., G. Liu, R.T. Qu, Z.F. Zhang, S.J. Wu, and T. Zhang. 2014. "Microstructural Percolation Assisted Breakthrough of Trade-Off Between Strength and Ductility in CuZr-Based Metallic Glass Composites." *Scientific Reports* 4, p. 4167.

Löser, W., J. Das, A. Guth, H.-J. Klauß, C. Mickel, U. Kühn, J. Eckert, S.K. Roy, and L. Schultz. 2004. "Effect of Casting Conditions on Dendrite-Amorphous/Nanocrystalline Zr–Nb–Cu–Ni–Al In Situ Composites." *Intermetallics* 12, no. 10–11, pp. 1153–8.

Lou, H.B., X.D. Wang, F. Xu, S.Q. Ding, Q.P. Cao, K. Hono, and J.Z. Jiang. 2011. "73 mm-Diameter Bulk Metallic Glass Rod by Copper Mould Casting." *Applied Physics Letters* 99, no. 5, p. 051910.

Louzguine-Luzgin, D.V., A. Vinogradov, G. Xie, S. Li, A. Lazarev, S. Hashimoto, and A. Inoue. 2009. "High-Strength and Ductile Glassy-Crystal Ni–Cu–Zr–Ti Composite Exhibiting Stress-Induced Martensitic Transformation." *Philosophical Magazine* 89, no. 32, pp. 2887–901.

Lowhaphandu, P. and J.J. Lewandowski. 1998. "Fracture Toughness and Notched Toughness of Bulk Amorphous Alloy: Zr–Ti–Ni–Cu–Be." *Scripta Materialia* 38, no. 12, pp. 1811–7.

Lowhaphandu, P., L.A. Ludrosky, S.L. Montgomery, and J.J. Lewandowski. 2000. "Deformation and Fracture Toughness of a Bulk Amorphous Zr–Ti–Ni–Cu–Be Alloy." *Intermetallics* 8, no. 5–6, pp. 487–92.

Lu, X.L., Q.H. Lu, Y. Li, and L. Lu. 2013. "Gradient Confinement Induced Uniform Tensile Ductility in Metallic Glass." *Scientific Reports* 3, p. 3319.

Lu, Z.P. and C.T. Liu. 2002. "A New Glass-Forming Ability Criterion for Bulk Metallic Glasses." *Acta Materialia* 50, no. 13, pp. 3501–12.

Lu, Z.P. and C.T. Liu. 2003. "Glass Formation Criterion for Various Glass-Forming Systems." *Physical Review Letters* 91, no. 11, p. 115505.

Lu, Z.P. and C.T. Liu. 2004. "A New Approach to Understanding and Measuring Glass Formation in Bulk Amorphous Materials." *Intermetallics* 12, no. 10–11, pp. 1035–43.

Lu, Z.P., Y. Liu, and C.T. Liu. 2008. "Evaluation of Glass-Forming Ability." In *Bulk Metallic Glasses*, ed. M. Miller and P. Liaw. Boston, MA: Springer, pp. 87–115.

Luo, J., H. Pan, and E.C. Kinzel. 2014. "Additive Manufacturing of Glass." *Journal of Manufacturing Science and Engineering* 136, no. 6, p. 061024.

Ma, E. 2015. "Tuning Order in Disorder." *Nature Materials* 14, no. 6, pp. 547–52.

Ma, D., H. Tan, Y. Zhang, and Y. Li. 2003. "Correlation between Glass Formation and Type of Eutectic Coupled Zone in Eutectic Alloys." *Materials Transactions* 44, no. 10, pp. 2007–10.

Ma, F., J. Yang, X. Zhu, C. Liang, and H. Wang. 2010. "Femtosecond Laser-Induced Concentric Ring Microstructures on Zr-Based Metallic Glass." *Applied Surface Science* 256, no. 11, pp. 3653–60.

Ma, H., Q. Zheng, J. Xu, and Y. Li. 2011. "Doubling the Critical Size for Bulk Metallic Glass Formation in the Mg–Cu–Y Ternary System." *Journal of Materials Research* 20, no. 9, pp. 2252–5.

Magagnosc, D.J., R. Ehrbar, G. Kumar, M.R. He, J. Schroers, and D.S. Gianola. 2013. "Tunable Tensile Ductility in Metallic Glasses." *Scientific Reports* 3, p. 1096.

Mani, M., B.M. Lane, M.A. Donmez, S.C. Feng, S.P. Moylan, and R.R. Fesperman, Jr. 2015. *Measurement Science Needs for Real-Time Control of Additive Manufacturing Powder Bed Fusion Processes*. Gaithersburg, MD: National Institute of Standards and Technology, NIST Interagency/Internal Report (NISTIR) 8036.

Marion, G., G. Cailletaud, C. Colin, and M. Maziere. 2014. *A Finite Element Model for the simulation of Direct Metal Deposition*. San Diego, CA: ICALEO.

Marion, G., G. Cailletaud, M. Mazière, C. Colin, and B. Missoum. 2015. Modelling additive manufacturing processes: from direct metal deposition to selective laser melting. *S09a Procédés de mise en forme*.

Markl, M. and C. Körner. 2016. "Multiscale Modeling of Powder Bed-Based Additive Manufacturing." *Annual Review of Materials Research* 46, no. 1, pp. 93–123.

Maruyama, S. 2000. "Molecular Dynamics Method for Microscale Heat Transfer." *Advances in Numerical Heat Transfer 2*, no. 6, pp. 189–226.

Maruyama, S. 2003. "A Molecular Dynamics Simulation of Heat Conduction of a Finite Length Single-Walled Carbon Nanotube." *Microscale Thermophysical Engineering* 7, no. 1, pp. 41–50.

Masrour, R. and A. Jabar. 2016. "Magnetic Properties of Dendrimer Structures with Different Coordination Numbers: A Monte Carlo Study." *Journal of Magnetism and Magnetic Materials* 417, pp. 397–400.

Matsumoto, M., M. Shiomi, K. Osakada, and F. Abe. 2002. "Finite Element Analysis of Single Layer Forming on Metallic Powder Bed in Rapid Prototyping by Selective Laser Processing." *International Journal of Machine Tools and Manufacture* 42, no. 1, pp. 61 67.

Mattern, N., P. Jovari, I. Kaban, S. Gruner, A. Elsner, V. Kokotin, H. Franz, B. Beuneu, and J. Eckert. 2009. "Short-Range Order of Cu–Zr Metallic Glasses." *Journal of Alloys and Compounds* 485, no. 1–2, pp. 163–9.

Matthews, D.T.A., V. Ocelík, D.Branagan, and J.Th.M. de Hosson. 2009. "Laser Engineered Surfaces From Glass Forming Alloy Powder Precursors: Microstructure and Wear." *Surface and Coatings Technology* 203, no. 13, pp. 1833–43.

Matthieu, M. 2016. "Relaxation and Physical Aging in Network Glasses: A Review." *Reports on Progress in Physics* 79, no. 6, p. 066504.

Maxwell, I. and A. Hellawell. 1975. "A Simple Model for Grain Refinement During Solidification." *Acta Metallurgica* 23, no. 2, pp. 229–37.

Mazzoli, A., G. Moriconi, and M.G. Pauri. 2007. "Characterization of an Aluminum-Filled Polyamide Powder for Applications in Selective Laser Sintering." *Materials & Design* 28, no. 3, pp. 993–1000.

Meiners, W., K.D. Wissenbach, and A.D. Gasser. 1998. "Shaped Body Especially Prototype or Replacement Part Production." *Google Patents.*

Metropolis, N., A.W. Rosenbluth, M.N. Rosenbluth, A.H. Teller, and E. Teller. 1953. "Equation of State Calculations by Fast Computing Machines." *The Journal of Chemical Physics* 21, no. 6, pp. 1087–92.

Michalik, S., J. Michalikova, M. Pavlovic, P. Sovak, H.-P. Liermann, and M. Miglierini. 2014. "Structural Modifications of Swift-Ion-Bombarded Metallic Glasses Studied by High-Energy X-Ray Synchrotron Radiation." *Acta Materialia* 80, pp. 309–16.

Miller, M.K. and P. Liaw. 2007. *Bulk Metallic Glasses: An Overview.* Boston, MA: Springer.

Miracle, D.B., D.V.L. Luzgin, L.V.L. Luzgin, and A. Inoue. 2010. "An Assessment of Binary Metallic Glasses: Correlations Between Structure, Glass Forming Ability and Stability." *International Materials Reviews* 55, no. 4, pp. 218–56.

Miracle, D.B., J.D. Miller, O.N. Senkov, C. Woodward, M.D. Uchic, and J. Tiely. 2014. "Exploration and Development of High Entropy Alloys for Structural Applications." *Entropy* 16, no. 1, pp. 494.

Modest, M.F. 2003. "Backward Monte Carlo Simulations in Radiative Heat Transfer." *Journal of Heat Transfer* 125, no. 1, pp. 57–62.

Mu, J., H. Fu, Z. Zhu, A. Wang, H. Li, Z. Hu, and H. Zhang. 2009. "Synthesis and Properties of Al–Ni–La Bulk Metallic Glass." *Advanced Engineering Materials* 11, no. 7, pp. 530–2.

Mu, J., Z. Zhu, R. Su, Y. Wang, H. Zhang, and Y. Ren. 2013. "In Situ High-Energy X-Ray Diffraction Studies of Deformation-Induced Phase Transformation in Ti-Based Amorphous Alloy Composites Containing Ductile Dendrites." *Acta Materialia* 61, no. 13, pp. 5008–17.

Mukherjee, S., Z. Zhou, J. Schroers, W.L. Johnson, and W.K. Rhim. 2004. "Overheating Threshold and Its Effect on Time–Temperature–Transformation Diagrams of Zirconium Based Bulk Metallic Glasses." *Applied Physics Letters* 84, no. 24, pp. 5010–2.

Murr, L.E. 2015. "Metallurgy of Additive Manufacturing: Examples From Electron Beam Melting." *Additive Manufacturing* 5, pp. 40–53.

Nandi, U.K., A. Banerjee, S. Chakrabarty, and S.M. Bhattacharyya. 2016. "Composition Dependence of the Glass Forming Ability in Binary Mixtures: The Role of Demixing Entropy." *The Journal of Chemical Physics* 145, no. 3, p. 034503.

Nelson, D.R. 1983. "Order, Frustration, and Defects in Liquids and Glasses." *Physical Review B* 28, no. 10, pp. 5515–35.

Nestler, B. and A. Choudhury. 2011. "Phase-Field Modeling of Multi-Component Systems." *Current Opinion in Solid State and Materials Science* 15, no. 3, pp. 93–105.

Ni, J. and C. Beckermann. 1991. "A Volume-Averaged Two-Phase Model for Transport Phenomena During Solidification." *Metallurgical Transactions B* 22, no. 3, pp. 349–61.

Nielsen, C.P. and H. Bruus. 2015. "Sharp-Interface Model of Electrodeposition and Ramified Growth." *Physical Review E* 92, no. 4, p. 042302.

Nishida, M., S. Li, K. Kitamura, T. Furukawa, A. Chiba, T. Hara, and K. Hiraga. 1998. "New Deformation Twinning Mode of B19′ Martensite in Ti–Ni Shape Memory Alloy." *Scripta Materialia* 39, no. 12, pp. 1749–54.

Nishiyama, N., K. Takenaka, H. Miura, N. Saidoh, Y. Zeng, and A. Inoue. 2012. "The World's Biggest Glassy Alloy Ever Made." *Intermetallics* 30, pp. 19–24.

Ocelík, V., N. Janssen, S.N. Smith, and J.Th.M. De Hosson. 2016. "Additive Manufacturing of High-Entropy Alloys by Laser Processing." *JOM* 68, no. 7, pp. 1810–8.

Ogata, S., F. Shimizu, J. Li, M. Wakeda, and Y. Shibutani. 2006. "Atomistic Simulation of Shear Localization in Cu–Zr Bulk Metallic Glass." *Intermetallics* 14, no. 8–9, pp. 1033–7.

Oh, Y.S., C.P. Kim, S. Lee, and N.J. Kim. 2011. "Microstructure and Tensile Properties of High-Strength High-Ductility Ti-Based Amorphous Matrix Composites Containing Ductile Dendrites." *Acta Materialia* 59, no. 19, pp. 7277–86.

Oh, J.C., T., Ohkobo, Y.C. Kim, E. Fluery, and K. Hono. 2005. "Phase Separation in Cu43Zr43Al7Ag7 Bulk Metallic Glass." *Scripta Materialia* 53, no. 2, pp. 165–9.

Olakanmi, E.O., R.F. Cochrane, and K.W. Dalgarno. 2015. "A Review on Selective Laser Sintering/Melting (SLS/SLM) of Aluminium Alloy Powders: Processing, Microstructure, and Properties." *Progress in Materials Science* 74, pp. 401–77.

Ott, R.T., F. Sansoz, T. Jiao, D. Warner, J.F. Molinari, K.T. Ramesh, T.C. Hufnagel, and C. Fan. 2006. "Yield Criteria and Strain-Rate Behavior of Zr57. 4Cu16.4Ni8.2Ta8Al10 Metallic–Glass–Matrix Composites." *Metallurgical and Materials Transactions A* 37, no. 11, pp. 3251–8.

Ott, R.T., F. Sansoz, J.F. Molinari, J. Almer, K.T. Ramesh, and T.C. Hufnagel. 2005. "Micromechanics of Deformation of Metallic–Glass–Matrix Composites From In Situ Synchrotron Strain Measurements and Finite Element Modeling." *Acta Materialia* 53, no. 7, pp. 1883–93.

Packard, C.E. and C.A. Schuh. 2007. "Initiation of Shear Bands Near a Stress Concentration in Metallic glass." *Acta Materialia* 55, no. 16, pp. 5348–58.

Pampillo, C.A. 1972. "Localized Shear Deformation in a Glassy Metal." *Scripta Metallurgica* 6, no. 10, pp. 915–7.

Pan, D., A. Inoue, T. Sakurai, and M.W. Chen. 2008. "Experimental Characterization of Shear Transformation Zones for Plastic Flow of Bulk Metallic Glasses." *Proceedings of the National Academy of Sciences* 105, no. 39, pp. 14769–72.

Paradis, P.-F., T. Ishikawa, G.W. Lee, D.H. Moritz, J. Brillo, W.K. Rhim, and J.T. Okada. 2014. "Materials Properties Measurements and Particle Beam Interactions Studies Using Electrostatic Levitation." *Materials Science and Engineering: R: Reports* 76, pp. 1–53.

Park, B.J., H.J. Chang, D.H. Kim, W.T. Kim, K. Chattopadhyay, T.A. Abinandanan, and S. Bhattacharyya. 2006. "Phase Separating Bulk Metallic Glass: A Hierarchical Composite." *Physical Review Letters* 96, no. 24, p. 245503.

Park, E.S., H.J. Chang, and D.H. Kim. 2008. "Effect of Addition of Be on Glass-Forming Ability, Plasticity and Structural Change in Cu–Zr Bulk Metallic Glasses." *Acta Materialia* 56, no. 13, pp. 3120–31.

Park, E.S. and D.H. Kim. 2005. "Design of Bulk Metallic Glasses With High Glass Forming Ability and Enhancement of Plasticity in Metallic Glass Matrix Composites: A Review." *Metals and Materials International* 11, no. 1, pp. 19–27.

Park, E.S. and D.H. Kim. 2006. "Phase Separation and Enhancement of Plasticity in Cu–Zr–Al–Y Bulk Metallic Glasses." *Acta Materialia* 54, no. 10, pp. 2597–604.

Park, E.S., J.S. Kyeong, and D.H. Kim. 2007. "Phase Separation and Improved Plasticity by Modulated Heterogeneity in Cu–(Zr, Hf)–(Gd, Y)–Al Metallic Glasses." *Scripta Materialia* 57, no. 1, pp. 49–52.

Park, J.M., N. Mattern, U. Kuhn, and J. Eckert. 2009. "High-Strength Bulk Al-Based Bimodal Ultrafine Eutectic Composite with Enhanced Plasticity." *Journal of Materials Research* 24, no. 08, pp. 2605–9.

Parr, R.G., D.P. Craig, and I.G. Ross. 1950. "Molecular Orbital Calculations of the Lower Excited Electronic Levels of Benzene, Configuration Interaction Included." *The Journal of Chemical Physics* 18, no. 12, pp. 1561–3.

Pauly, S., J. Das, J. Bednarcik, N. Mattern, K.B. Kim, D.H. Kim, and J. Eckert. 2009a. "Deformation-Induced Martensitic Transformation in Cu–Zr–(Al,Ti) Bulk Metallic Glass Composites." *Scripta Materialia* 60, no. 6, pp. 431–4.

Pauly, S., S. Gorantla, G. Wang, U. Kuhn, and J. Eckert. 2010. "Transformation-Mediated Ductility in CuZr-Based Bulk Metallic Glasses." *Nature Materials* 9, no. 6, pp. 473–7.

Pauly, S., C. Liu, G. Wang, J. Das, K.B. Kim, U. Kuhn, D.H. Kim, and J. Eckert. 2009b. "Modeling Deformation Behavior of Cu–Zr–Al Bulk Metallic Glass Matrix Composites." *Applied Physics Letters* 95, no. 10, p. 101906.

Pauly, S., L. Lober, R. Petters, M. Stoica, S. Scudino, U. Kuhn, and J. Eckert. 2013. "Processing Metallic Glasses by Selective Laser Melting." *Materials Today* 16, no. 1–2, pp. 37–41.

Pekarskaya, E., C.P. Kim, and W.L. Johnson. 2001. "In Situ Transmission Electron Microscopy Studies of Shear Bands in a Bulk Metallic Glass Based Composite." *Journal of Materials Research* 16, no. 09, pp. 2513–8.

Peker, A. and W.L. Johnson. 1993. "A Highly Processable Metallic Glass: Zr41. 2Ti13.8Cu12.5Ni10.0Be22.5." *Applied Physics Letters* 63, no. 17, pp. 2342–4.

Perim, E., D. Lee, Y. Liu, C. Toher, P. Gong, Y. Li, N. Simmons, et al. 2016. "Spectral Descriptors for Bulk Metallic Glasses Based on the Thermodynamics of Competing Crystalline Phases." *Nature Communications* 7, p. 12315.

Pinkerton, A.J. 2016. "[INVITED] Lasers in Additive Manufacturing." *Optics & Laser Technology* 78, Part A, pp. 25–32.

Planas Almazan, P. 1997. Accuracy of Monte Carlo ray-tracing thermal radiation calculations: A practical discussion. In *Sixth European Symposium on Space Environmental Control Systems*. European Space Agency, SP-400, Noordwijk, The Netherlands, 20–22 May, 1997.

Porter, D.A. and K.E. Easterling. 1992. *Phase Transformations in Metals and Alloys*. 3rd ed. (Revised Reprint). London: Taylor & Francis.

Prashanth, K.G., H.S. Shahabi, H. Attar, V.C. Srivastava, N. Ellendt, V. Uhlenwinkel, J. Eckert, and S. Scudino. 2015. "Production of High Strength Al85Nd8Ni5Co2 Alloy by Selective Laser Melting." *Additive Manufacturing* 6, pp. 1–5.

Price, C.W. 1987. "Simulations of Grain Impingement and Recrystallization Kinetics." *Acta Metallurgica* 35, no. 6, pp. 1377–90.

Pronk, S., S. Pall, R. Schulz, P. Larsson, P. Bjelkmar, R. Apostolov, M.R. Shirts, et al. 2013. "GROMACS 4.5: A High-Throughput and Highly Parallel Open Source Molecular Simulation Toolkit." *Bioinformatics* 29, no. 7, pp. 845–54.

Pusztai, L. and E. Sváb. 1993. "Structure Study of Ni62Nb38 Metallic Glass Using Reverse Monte Carlo Simulation." *Journal of Non-Crystalline Solids* 156, pp. 973–7.

Qian, M. 2015. "Metal Powder for Additive Manufacturing." *JOM* 67, no. 3, pp. 536–7.

Qiao, J. 2013. "In-Situ Dendrite/Metallic Glass Matrix Composites: A Review." *Journal of Materials Science & Technology* 29, no. 8, pp. 685–701.

Qiao, J., H. Jia, and P.K. Liaw. 2016. "Metallic Glass Matrix Composites." *Materials Science and Engineering: R: Reports* 100, pp. 1–69.

Qiao, J.W., E.W. Huang, G.Y. Wang, H.J. Yang, W. Liang, Y. Zhang, and P.K. Liaw. 2013a. "Characteristic of Improved Fatigue Performance for Zr-Based Bulk Metallic Glass Matrix Composites." *Materials Science and Engineering: A* 563, pp. 101–5.

Qiao, J.C. and J.M. Pelletier. 2014. "Dynamic Mechanical Relaxation in Bulk Metallic Glasses: A Review." *Journal of Materials Science & Technology* 30, no. 6, pp. 523–45.

Qiao, J.W., H.Y. Ye, Y.S. Wang, S. Pauly, H.J. Yang, and Z.H. Wang. 2013b. "Distinguished Work-Hardening Capacity of a Ti-Based Metallic Glass Matrix Composite Upon Dynamic Loading." *Materials Science and Engineering: A* 585, pp. 277–80.

Qiao, J.W., H.Y. Ye, H.J. Yang, W. Liang, B.S. Xu, P.K. Liaw, and M.W. Chen. 2013c. "Dynamic Shear Punching of Metallic Glass Matrix Composites." *Intermetallics* 36, pp. 31–35.

Qiao, J.W., Y. Zhang, and P.K. Liaw. 2010. "Serrated Flow Kinetics in a Zr-Based Bulk Metallic Glass." *Intermetallics* 18, no. 11, pp. 2057–64.

Qiao, J.W., Y. Zhang, P.K. Liaw, and G.L. Chen. 2009a. "Micromechanisms of Plastic Deformation of a Dendrite/Zr-Based Bulk–Metallic–Glass Composite." *Scripta Materialia* 61, no. 11, pp. 1087–90.

Qiao, J.W., Y. Zhang, Z.L. Zheng, J.P. He, and B.C. Wei. 2009b. "Synthesis of Plastic Zr-Based Bulk Metallic Glass Matrix Composites by the Copper-Mould

Suction Casting and the Bridgman Solidification." *Journal of Alloys and Compounds* 477, no. 1–2, pp. 436–9.

Qu, R.T., M. Calin, J. Eckert, and Z.F. Zhang. 2012. "Metallic Glasses: Notch-Insensitive Materials." *Scripta Materialia* 66, no. 10, pp. 733–6.

Quadrini, F. and L. Santo. 2008. "Selective Laser Sintering of Resin-Coated Sands—Part I: The Laser-Material Interaction." *Journal of Manufacturing Science and Engineering* 131, no. 1, p. 011004.

Quintana, I., T. Dobrev, A. Aranzabe, G. Lalev, and S. Dimov. 2009. "Investigation of Amorphous and Crystalline Ni Alloys Response to Machining With Micro-Second and Pico-Second Lasers." *Applied Surface Science* 255, no. 13–14, pp. 6641–6.

Raabe, D. 2004. "Overview of the Lattice Boltzmann Method for Nano- and Microscale Fluid Dynamics in Materials Science and Engineering." *Modelling and Simulation in Materials Science and Engineering* 12, no. 6, p. R13.

Rafique, M.M.A. 2015. "Modeling and Simulation of Heat Transfer Phenomena." In *Heat Transfer*, ed. Salim Newaz Kazi. Vienna, Austria: Intech Publishers. DOI: 10.5772/61029.

Rafique, M.M.A. and J. Iqbal. 2009. "Modeling and Simulation of Heat Transfer Phenomena During Investment Casting." *International Journal of Heat and Mass Transfer* 52, no. 7–8, pp. 2132–9.

Ramil, A., J. Lamas, J.C. Alvarez, A.J. Lopez, E. Saavedra, and A. Yanez. 2009. "Micromachining of Glass by the Third Harmonic of Nanosecond Nd:YVO4 Laser." *Applied Surface Science* 255, no. 10, pp. 5557–60.

Rappaz, M. and E. Blank. 1986. "Simulation of Oriented Dendritic Microstructures Using the Concept of Dendritic Lattice." *Journal of Crystal Growth* 74, no. 1, pp. 67–76.

Rappaz, M. and C.A. Gandin. 1993. "Probabilistic Modelling of Microstructure Formation in Solidification Processes." *Acta Metallurgica et Materialia* 41, no. 2, pp. 345–60.

Rappaz, M. 1989. "Modelling of Microstructure Formation in Solidification Processes." *International Materials Reviews* 34, no. 1, pp. 93–124.

Ray, C.S., S.T. Reis, R.K. Brow, W. Holand, and V. Rheinberger. 2005. "A New DTA Method for Measuring Critical Cooling Rate for Glass Formation." *Journal of Non-Crystalline Solids* 351, no. 16–17, pp. 1350–8.

Rim, K.R., J.M. Park, W.T. Kim, and D.H. Kim. 2013. "Tensile Necking and Enhanced Plasticity of Cold Rolled β–Ti Dendrite Reinforced Ti-Based Bulk Metallic Glass Matrix Composite." *Journal of Alloys and Compounds* 579, pp. 253–8.

Ritchie, R.O. 2011. "The Conflicts Between Strength and Toughness." *Nature Materials* 10, no. 11, pp. 817–22.

Romano, J., L. Ladani, J. Razmi, and M. Sadowski. 2015. "Temperature Distribution and Melt Geometry in Laser and Electron-Beam Melting Processes – A Comparison Among Common Materials." *Additive Manufacturing* 8, pp. 1–11.

Rombouts, M. 2006. *Selective Laser Sintering/Melting of Iron-Based Powders (Selectief laser sinteren/smelten van ijzergebaseerde poeders)* [PhD Thesis]. Leuven, Belgium: Department of Metallrgy, Faculty of Engineering, Katholic University.

Sames, W.J., F.A. List, S. Pannala, R.R. Dehoff, and S.S. Babu. 2016. "The Metallurgy and Processing Science of Metal Additive Manufacturing." *International Materials Reviews* 61, no. 5, pp. 315–60.

Santos, E.C., M. Shiomi, K. Osakada, and T. Laoui. 2006. "Rapid Manufacturing of Metal Components by Laser Forming." *International Journal of Machine Tools and Manufacture* 46, no. 12–13, pp. 1459–68.

Sarac, B., G. Kumar, T. Hodges, S. Ding, A. Desai, and J. Schroers. 2011. "Three-Dimensional Shell Fabrication Using Blow Molding of Bulk Metallic Glass." *Journal of Microelectromechanical Systems* 20, no. 1, pp. 28–36.

Sarac, B. and J. Schroers. 2013. "Designing Tensile Ductility in Metallic Glasses." *Nature Communications* 4, p. 2158.

Sato, Y. and B. Ničeno. 2013. "A Sharp-Interface Phase Change Model for a Mass-Conservative Interface Tracking Method." *Journal of Computational Physics* 249, pp. 127–61.

Schmidt, M., M. Zaeh, T. Graf, and A. Ostendorf. 2011. "Lasers in Manufacturing 2011 – Proceedings of the Sixth International WLT Conference on Lasers in Manufacturing the Property Research on High-Entropy Alloy AlxFeCoNiCuCr Coating by Laser Cladding." *Physics Procedia* 12, pp. 303–12.

Schroers, J. 2005. "The Superplastic Forming of Bulk Metallic Glasses." *JOM* 57, no. 5, pp. 35–39.

Schroers, J. 2010. "Processing of Bulk Metallic Glass." *Advanced Materials* 22, no. 14, pp. 1566–97.

Schroers, J. and W.L. Johnson. 2004. "Ductile Bulk Metallic Glass." *Physical Review Letters* 93, no. 25, p. 255506.

Schryvers, D., G.S. Firstov, J.W. Seo, J.V Humbeeck, and Y.N. Koval. 1997. "Unit Cell Determination in CuZr Martensite by Electron Microscopy and X-Ray Diffraction." *Scripta Materialia* 36, no. 10, pp. 1119–25.

Schryvers, D., P. Potapov, R. Santamarta, and W. Tirry. 2004. "Applications of Advanced Transmission Electron Microscopic Techniques to Ni–Ti Based Shape Memory Materials." *Materials Science and Engineering: A* 378, no. 1–2, pp. 11–15.

Schuh, C.A., T.C. Hufnagel, and U. Ramamurty. 2007. "Mechanical Behavior of Amorphous Alloys." *Acta Materialia* 55, no. 12, pp. 4067–109.

Schulz, R., B. Lindner, L. Petridis, and J.C. Smith. 2009. "Scaling of Multimillion-Atom Biological Molecular Dynamics Simulation on a Petascale Supercomputer." *Journal of Chemical Theory and Computation* 5, no. 10, pp. 2798–808.

Seabaugh, A. 2013. "The Tunneling Transistor." *IEEE Spectrum* 50, no. 10, pp. 35–62.

Seo, J.W. and D. Schryvers. 1998a. "TEM Investigation of the Microstructure and Defects of CuZr Martensite. Part I: Morphology and Twin Systems." *Acta Materialia* 46, no. 4, pp. 1165–75.

Seo, J.W. and D. Schryvers. 1998b. "TEM Investigation of the Microstructure and Defects of CuZr Martensite. Part II: Planar Defects." *Acta Materialia* 46, no. 4, pp. 1177–83.

Sha, Z.D. and Q.X. Pei. 2015. "Ab Initio Study on the Electronic Origin of Glass-Forming Ability in the Binary Cu–Zr and the Ternary Cu–Zr–Al(Ag) Metallic Glasses." *Journal of Alloys and Compounds* 619, pp. 16–19.

Shamsaei, N., A. Yadollahi, L. Bian, and L.M. Thompson. 2015. "An Overview of Direct Laser Deposition for Additive Manufacturing; Part II: Mechanical Behavior, Process Parameter Optimization and Control." *Additive Manufacturing* 8, pp. 12–35.

Shen, T.D., B.R. Sun, and S.W. Xin. 2015. "Effects of Metalloids on the Thermal Stability and Glass Forming Ability of Bulk Ferromagnetic Metallic Glasses." *Journal of Alloys and Compounds* 631, pp. 60–66.

Sheng, H.W., W.K. Luo, F.M. Alamgir, J.M. Bai, and E. Ma. 2006. "Atomic Packing and Short-to-Medium-Range Order in Metallic Glasses." *Nature* 439, no. 7075, pp. 419–25.

Shi, Y. and M.L. Falk. 2006. "Does Metallic Glass have a Backbone? The Role of Percolating Short Range Order in Strength and Failure." *Scripta Materialia* 54, no. 3, pp. 381–6.

Shi, Y. and M.L. Falk. 2007. "Stress-Induced Structural Transformation and Shear Banding During Simulated Nanoindentation of a Metallic Glass." *Acta Materialia* 55, no. 13, pp. 4317–24.

Shifeng, W., L. Shuai, W. Qingsong, C. Yan, Z. Sheng, and S. Yusheng. 2014. "Effect of Molten Pool Boundaries on the Mechanical Properties of Selective Laser Melting Parts." *Journal of Materials Processing Technology* 214, no. 11, pp. 2660–7.

Shinno, S. and U.K. Yamada新野, 俊. and 英. 山田. 2009. 粉末焼結積層造形による透明部品の作製 ―屈折率調整されたエポキシ樹脂の含浸によるプラスチック粉末焼結積層造形品の透明化―. 精密工学会誌 75, no. 12, pp. 1454–8.

Shiomi, M., A. Yoshidome, F. Abe, and K. Osakada. 1999. "Finite Element Analysis of Melting and Solidifying Processes in Laser Rapid Prototyping of Metallic Powders." *International Journal of Machine Tools and Manufacture* 39, no. 2, pp. 237–52.

Shishkovsky, I., I. Yadroitsev, Ph. Bertrand, and I. Smurov. 2007. "Alumina–Zirconium Ceramics Synthesis by Selective Laser Sintering/Melting." *Applied Surface Science* 254, no. 4, pp. 966–70.

Shukla, M. and V. Verma. 2014. Finite element simulation and analysis of laser metal deposition. In *6th International Conference on Mechanical, Production and Automobile Engineering*, November 27–28. Cape Town, South Africa.

Sigl, M., S. Lutzmann, and M. Zäh. 2006. Transient physical effects in electron beam sintering. In *Solid Freeform Fabrication Symposium Proceedings*, Austin, TX.

Smith, J., W. Xiong, W. Yan, S. Lin, P. Cheng, O.L. Kafka, G.J. Wagner, J. Cao, and W.K. Liu. 2016. "Linking Process, Structure, Property, and Performance for Metal-Based Additive Manufacturing: Computational Approaches With Experimental Support." *Computational Mechanics* 57, no. 4, pp. 583–610.

Smugeresky, J., D.M. Keicher, J.A. Romero, M.L. Griffith, and L.D. Harwell. 1997. "Laser Engineered Net Shaping (LENS) Process: Optimization of Surface Finish and Microstructural Properties." *Advances in Powder Metallurgy and Particulate Materials* 3, p. 21.

Snook, I. 2006. *The Langevin and Generalised Langevin Approach to the Dynamics Of Atomic, Polymeric and Colloidal Systems*. Amsterdam, The Netherlands: Elsevier.

Song, K. 2013. *Synthesis, Microstructure, and Deformation Mechanisms of CuZr-Based Bulk Metallic Glass Composites* [dissertation]. Germany: Technische Universitat Dresdon.

Song, K.K., S. Pauly, B.A. Sun, J. Tan, M. Stoica, U. Kuhn, and J. Eckert. 2013. "Correlation Between the Microstructures and the Deformation Mechanisms of CuZr-Based Bulk Metallic Glass Composites." *AIP Advances* 3, no. 1, p. 012116.

Song, K.K., S. Pauly, Y. Zhang, P. Gargarella, R. Li, N.S. Barekar, U. Kuhn, M. Stoica, and J. Eckert. 2011. "Strategy for Pinpointing the Formation of B2 CuZr in Metastable CuZr-Based Shape Memory Alloys." *Acta Materialia* 59, no. 17, pp. 6620–30.

Song, K.K., S. Pauly, Y. Zhang, R. Li, S. Gorantla, N. Narayanan, U. Kuhn, T. Gemming, and J. Eckert. 2012. "Triple Yielding and Deformation Mechanisms in Metastable Cu47.5Zr47.5Al5 Composites." *Acta Materialia* 60, no. 17, pp. 6000–12.

Song, H., R. Shi, Y. Wang, and J.J. Hoyt. 2016. "Simulation Study of Heterogeneous Nucleation at Grain Boundaries During the Austenite-Ferrite Phase Transformation: Comparing the Classical Model With the Multi-Phase Field Nudged Elastic Band Method." *Metallurgical and Materials Transactions A*, pp. 1–9.

Song, K.J., Y.H. Wei, Z.B. Dong, R. Ma, X.H. Zhan, W.J. Zheng, and K. Fang. 2014. "Virtual Front Tracking Cellular Automaton Modeling of Isothermal β to α Phase Transformation With Crystallography Preferred Orientation of TA15 Alloy." *Modelling and Simulation in Materials Science and Engineering*, 22, no. 1, p. 015006.

Song, W., Y. Wu, H. Wang, X. Liu, H. Chen, Z. Guo, and Z. Lu. 2016. "Microstructural Control via Copious Nucleation Manipulated by In Situ Formed Nucleants: Large-Sized and Ductile Metallic Glass Composites." *Advanced Materials* 28, no. 37, pp. 8156–61.

Spaepen, F. 1977. "A Microscopic Mechanism for Steady State Inhomogeneous Flow in Metallic Glasses." *Acta Metallurgica* 25, no. 4, pp. 407–15.

Spears, T.G. and S.A. Gold. 2016. "In-Process Sensing in Selective Laser Melting (SLM) Additive Manufacturing." *Integrating Materials and Manufacturing Innovation* 5, no. 1, pp. 1–25.

Spittle, J.A. and S.G.R. Brown. 1989. "Computer Simulation of the Effects of Alloy Variables on the Grain Structures of Castings." *Acta Metallurgica*, 37, no. 7, pp. 1803–10.

Stefanescu, D. 2015. *Science and Engineering of Casting Solidification*. Dordrecht: Springer.

Steif, P.S., F. Spaepen, and J.W. Hutchinson. 1982. "Strain Localization in Amorphous Metals." *Acta Metallurgica* 30, no. 2, pp. 447–55.

Steinhardt, P.J. 1990. "Quasicrystals: A New Form of Matter." Endeavour 14, no. 3, pp. 112–6.

StJohn, D.H., M. Qian, M.A. Easton, and P. Cao. 2011. "The Interdependence Theory: The Relationship Between Grain Formation and Nucleant Selection." *Acta Materialia* 59, no. 12, pp. 4907–21.

Stoica, M., J. Das, J. Bednarok, G. Wang, G. Vaughan, W.H. Wang, and J. Eckert. 2010. "Mechanical Response of Metallic Glasses: Insights From In-Situ High Energy X-Ray Diffraction." *JOM* 62, no. 2, pp. 76–82.

Sun, B.A., S. Pauly, J. Hu, W.H. Wang, U. Kuhn, and J. Eckert. 2013. "Origin of Intermittent Plastic Flow and Instability of Shear Band Sliding in Bulk Metallic Glasses." *Physical Review Letters* 110, no. 22, p. 225501.

Sun, B.A., S. Pauly, J. Tan, M. Stoica, W.H. Wang, U. Kuhn, and J. Eckert. 2012. "Serrated Flow and Stick–Slip Deformation Dynamics in the Presence of Shear-Band Interactions for a Zr-Based Metallic Glass." *Acta Materialia* 60, no. 10, pp. 4160–71.

Sun, D., M. Zhu, S. Pan, and D. Raabe. 2009. "Lattice Boltzmann Modeling of Dendritic Growth in a Forced Melt Convection." *Acta Materialia* 57, no. 6, pp. 1755–67.

Sun, D.K., M.-F. Zhu, T. Dai, W.-S. Cao, S.-L. Chen, D. Raabe, and C.-P. Hong. 2011a. "Modelling of Dendritic Growth in Ternary Alloy Solidification With Melt Convection." *International Journal of Cast Metals Research* 24, no. 3–4, pp. 177–83.

Sun, D.K., M.F. Zhu, S.Y. Pan, C.R. Yang, and D. Raabe. 2011b. "Lattice Boltzmann Modeling of Dendritic Growth in Forced and Natural Convection." *Computers & Mathematics with Applications* 61, no. 12, pp. 3585–92.

Sun, H. and K.M. Flores. 2008. "Laser Deposition of a Cu-Based Metallic Glass Powder on a Zr-Based Glass Substrate." *Journal of Materials Research* 23, no. 10, pp. 2692–703.

Sun, H. and K.M. Flores. 2010. "Microstructural Analysis of a Laser-Processed Zr-Based Bulk Metallic Glass." *Metallurgical and Materials Transactions A* 41, no. 7, pp. 1752–57.

Sun, H. and K.M. Flores. 2013. "Spherulitic Crystallization Mechanism of a Zr-Based Bulk Metallic Glass During Laser Processing." *Intermetallics* 43, pp. 53–59.

Sun, L., J. Wang, H. Kou, J. Li, and P. Zhang. 2016. "Phase Separation and Microstructure Evolution of Zr48Cu36Ag8Al8 Bulk Metallic Glass in the Supercooled Liquid Region." *Rare Metal Materials and Engineering* 45, no. 3, pp. 567–70.

Sun, S.-H., Y. Koizumi, S. Kurosu, Y.-P. Li, H. Matsumoto, and A. Chiba. 2014. "Build Direction Dependence of Microstructure and High-Temperature Tensile Property of Co–Cr–Mo Alloy Fabricated by Electron Beam Melting." *Acta Materialia* 64, pp. 154–68.

Sun, Y.F., T.L. Cheung, Y.R. Wang, C.H. Shek, W.H. Li, and B.C. Wei. 2005a. "Effect of Quasicrystalline Phase on the Deformation Behavior of Zr62Al9.5Ni9. 5Cu14Nb5 Bulk Metallic Glass." *Materials Science and Engineering: A* 398, no. 1–2, pp. 22–27.

Sun, Y.F., S.K. Guan, B.C. Wei, Y.R. Wang, and C.H. Shek. 2005b. "Brittleness of Zr-Based Bulk Metallic Glass Matrix Composites Containing Ductile Dendritic Phase." *Materials Science and Engineering: A* 406, no. 1–2, pp. 57–62.

Sun, Y.F., C.H. Shek, S.K. Guan, B.C. Wei, and J.Y. Geng. 2006. "Formation, Thermal Stability and Deformation Behavior of Graphite-Flakes Reinforced Cu-Based Bulk Metallic Glass Matrix Composites." *Materials Science and Engineering: A* 435–436, pp. 132–38.

Sun, Y.F., C.H. Shek, B.C. Wei, W.H. Li, and Y.R. Wang. 2005c. "Effect of Nb Content on the Microstructure and Mechanical Properties of Zr–Cu–Ni–Al–Nb Glass Forming Alloys." *Journal of Alloys and Compounds* 403, no. 1–2, pp. 239–244.

Taltavull, C., B. Torres, A.J. Lopez, P. Rodrigo, E. Otero, and J. Rams. 2012. "Selective Laser Surface Melting of a Magnesium–Aluminium Alloy." *Materials Letters* 85, pp. 98–101.

Tan, J., C.J. Li, Y.H. Jiang, R. Zhou, J. Eckert, 2013. "Correlation Between Internal States and Strength in Bulk Metallic Glass." In *PRICM*. John Wiley & Sons, Inc., pp. 3199–206.

Tan, J., Y. Zhang, M. Stoica, U. Kuhn, N. Mattern, F.S. Pan, and J. Eckert. 2011. "Study of Mechanical Property and Crystallization of a ZrCoAl Bulk Metallic Glass." *Intermetallics* 19, no. 4, pp. 567–71.

Tan, W., N.S. Bailey, and Y.C. Shin. 2011. "A Novel Integrated Model Combining Cellular Automata and Phase Field Methods for Microstructure Evolution During Solidification of Multi-Component and Multi-Phase Alloys." *Computational Materials Science* 50, no. 9, pp. 2573–85.

Tang, C. and C.H. Wong. 2015. "Effect of Atomic-Level Stresses on Local Dynamic and Mechanical Properties in CuxZr100−x Metallic Glasses: A Molecular Dynamics Study." *Intermetallics* 58, pp. 50–55.

Tang, H.P., G.Y. Yang, W.P. Jia, W.W. He, S.L. Lu, and M. Qian.2015. "Additive Manufacturing of a High Niobium-Containing Titanium Aluminide Alloy by Selective Electron Beam Melting." *Materials Science and Engineering: A* 636, pp. 103–7.

Tang, Y., H.T. Loh, Y.S. Wong, J.Y.H. Fuh, L. Lu, and X. Wang. 2003. "Direct Laser Sintering of a Copper-Based Alloy for Creating Three-Dimensional Metal Parts." *Journal of Materials Processing Technology* 140, no. 1–3, pp. 368–72.

Taub, A.I. and F. Spaepen. 1980. "The Kinetics of Structural Relaxation of a Metallic Glass." *Acta Metallurgica* 28, no. 12, pp. 1781–8.

Tauqir, A. 1986. *Transport Phenomena and Microstructural Developments During Electron Beam Melting (Rapid Solidification, Fluid Flow, Carbides, Tool Steel).*

Tejaswini, N., K.R. Babu, and K.S. Ram. 2015. Functionally graded material: An overview. In *International Conference on Advances in Engineering Science and Management.*

Telford, M. 2004. "The Case for Bulk Metallic Glass." *Materials Today* 7, no. 3, pp. 36–43.

Thévoz, P., J.L. Desbiolles, and M. Rappaz. 1989. "Modeling of Equiaxed Microstructure Formation in Casting." *Metallurgical Transactions A* 20, no. 2, pp. 311–22.

Thijs, L., F. Verhaeghe, T. Craeghs, J.V. Humbeeck, and J.P. Kruth. 2010. "A Study of the Microstructural Evolution During Selective Laser Melting of Ti–6Al–4V." *Acta Materialia* 58, no. 9, pp. 3303–12.

Thompson, S.M., L. Bian, N. Shamsaei, and A. Yadollahi. 2015. "An Overview of Direct Laser Deposition for Additive Manufacturing; Part I: Transport Phenomena, Modeling and Diagnostics." *Additive Manufacturing* 8, pp. 36–62.

Tian, F., Z. Li, and J. Song. 2016. "Solidification of Laser Deposition Shaping for TC4 Alloy Based on Cellular Automation." *Journal of Alloys and Compounds* 676, pp. 542–50.

Travitzky, N., A. Bonet, B. Dermeik, T. Fey, I. Fibert-Demut, L. Schlier, T. Schordt, and P. Greil. 2014. "Additive Manufacturing of Ceramic-Based Materials." *Advanced Engineering Materials* 16, no. 6, pp. 729–54.

Trexler, M.M. and N.N. Thadhani. 2010. "Mechanical Properties of Bulk Metallic Glasses." *Progress in Materials Science* 55, no. 8, pp. 759–839.

Trivedi, R., P. Magnin, and W. Kurz. 1987. "Theory of Eutectic Growth Under Rapid Solidification Conditions." *Acta Metallurgica* 35, no. 4, pp. 971–80.

Tsai, D.-C. and W.-S. Hwang. 2011. "A Three Dimensional Cellular Automaton Model for the Prediction of Solidification Morphologies of Brass Alloy by Horizontal Continuous Casting and Its Experimental Verification." *Materials Transactions* 52, no. 4, pp. 787–94.

Tsai, P. and K.M. Flores. 2015. "A Laser Deposition Strategy for the Efficient Identification of Glass-Forming Alloys." *Metallurgical and Materials Transactions A* 46, no. 9, pp. 3876–82.

Tsao, S.S. and F. Spaepen. 1985. "Structural Relaxation of a Metallic Glass Near Equilibrium." *Acta Metallurgica* 33, no. 5, pp. 881–9.

Turnbull, D. 1969. "Under What Conditions Can a Glass Be Formed?" *Contemporary Physics* 10, no. 5, pp. 473–88.

Van De Moortèle, B., T. Epicier, J.M. Pelletier, and J.L. Soubeyroux. 2004. "Phase Separation Before Crystallization in Zr–Ti–Cu–Ni–Be Bulk Metallic Glasses:

Influence of the Chemical Composition." *Journal of Non-Crystalline Solids* 345–346, pp. 169–72.

Van Elsen, M. 2007. *Complexity of Selective Laser Melting: A New Optimisation Approach* [doctoral dissertation]. Belgium: Katholic Universitat Leuveun.

Vandenbroucke, B. and J.P. Kruth. 2007. "Selective Laser Melting of Biocompatible Metals for Rapid Manufacturing of Medical Parts." *Rapid Prototyping Journal* 13, no. 4, pp. 196–203.

Vastola, G., G. Zhang, Q.X. Pei, and Y.-W. Zhang. 2016. "Modeling the Microstructure Evolution During Additive Manufacturing of Ti6Al4V: A Comparison Between Electron Beam Melting and Selective Laser Melting." *JOM* 68, no. 5, pp. 1370–5.

Vermolen, F.J. 2006. *Zener Solutions for Particle Growth in Multi-Component Alloys.* Delft University of Technology, Faculty of Electrical Engineering, Mathematics and Computer Science, Delft Institute of Applied Mathematics.

Vermolen, F., P. Pinson, C. Vuik, and S. van der Zwang. 2006. A suite of sharp interface models for solid-state transformations in alloys. In *ECCOMAS CFD 2006: Proceedings of the European Conference on Computational Fluid Dynamics, Egmond aan Zee*, The Netherlands, September 5–8, 2006. Delft University of Technology; European Community on Computational Methods in Applied Sciences (ECCOMAS).

Voller, V.R., A.D. Brent, and C. Prakash. 1989. "The Modelling of Heat, Mass and Solute Transport in Solidification Systems." *International Journal of Heat and Mass Transfer* 32, no. 9, pp. 1719–31.

Vutova, K. and V. Donchev. 2013. "Electron Beam Melting and Refining of Metals: Computational Modeling and Optimization." *Materials* 6, no. 10, p. 4626.

Wada, T., A. Inoue, and A.L. Greer. 2005. "Enhancement of Room-Temperature Plasticity in a Bulk Metallic Glass by Finely Dispersed Porosity." *Applied Physics Letters* 86, no. 25, p. 251907.

Wadhwa, P.P., J. Heinrich, and R. Busch. 2007. "Processing of Copper Fiber-Reinforced Zr41.2Ti13.8Cu12.5Ni10.0Be22.5 Bulk Metallic Glass Composites." Scripta Materialia 56, no. 1, pp. 73–76.

Wang, B., J.Y. Zhang, X.M. Li, and W.H. Qi. 2010a. "Simulation of Solidification Microstructure in Twin-Roll Casting Strip." *Computational Materials Science* 49, no. 1, Supplement, pp. S135–9.

Wang, C.Y., and C. Beckermann. 1993. "A Multiphase Solute Diffusion Model for Dendritic Alloy Solidification." *Metallurgical Transactions A* 24, no. 12, pp. 2787–802.

Wang, D., Y. Li, B.B. Sun, M.L. Sui, K. Lu, and E. Ma. 2004. "Bulk Metallic Glass Formation in the Binary Cu-Zr System." *Applied Physics Letters* 84, no. 20, pp. 4029–31.

Wang, D., G. Liao, J. Pan, Z. Tang, P. Peng, L. Liu, and T. Shi. 2009a. "Superplastic Micro-Forming of Zr65Cu17.5Ni10Al7.5 Bulk Metallic Glass With Silicon Mold Using Hot Embossing Technology." *Journal of Alloys and Compounds* 484, no. 1–2, pp. 118–22.

Wang, G., Y.J. Huang, M. Shagiev, and J. Shen. 2012a. "Laser Welding of Ti40Zr25Ni3Cu12Be20 Bulk Metallic Glass." *Materials Science and Engineering A* 541, pp. 33–37.

Wang, G., S. Pauly, S. Gorantla, N. Mattern, and J. Eckert. 2014a. "Plastic Flow of a Cu50Zr45Ti5 Bulk Metallic Glass Composite." *Journal of Materials Science & Technology* 30, no. 6, pp. 609–15.

Wang, G.Y., P.K. Liaw, and M.L. Morrison. 2009b. "Progress in Studying the Fatigue Behavior of Zr-Based Bulk-Metallic Glasses and Their Composites." *Intermetallics* 17, no. 8, pp. 579–90.

Wang, H.-S., H.-G. Chen, and J.S.-C. Jang. 2010. "Microstructure Evolution in Nd:YAG Laser-Welded (Zr53Cu30Ni9Al8)Si0.5 Bulk Metallic Glass Alloy." *Journal of Alloys and Compounds* 495, no. 1, pp. 224–8.

Wang, H.S., H.G. Chen, J.S.C. Jang, and M.S. Chiou. 2010b. "Combination of a Nd:YAG Laser and a Liquid Cooling Device to (Zr53Cu30Ni9Al8)Si0.5 Bulk Metallic Glass Welding." *Materials Science and Engineering A* 528, no. 1, pp. 338–41.

Wang, H.-S., M.S. Chiou, H.G. Chen, and J.S.-C. Jang. 2011. "The Effects of Initial Welding Temperature and Welding Parameters on the Crystallization Behaviors of Laser Spot Welded Zr-Based Bulk Metallic Glass." *Materials Chemistry and Physics* 129, no. 1–2, pp. 547–52.

Wall, J.J., C. Fan, P.K. Liaw, C.T. Liu, and H. Choo. 2006. "A Combined Drop/Suction-Casting Machine for the Manufacture of Bulk-Metallic-Glass Materials." *Review of Scientific Instruments* 77, no. 3, p. 033902.

Wang, L.-M., Y. Tian, R. Liu, and W. Wang. 2012b. "A 'universal' Criterion for Metallic Glass Formation." *Applied Physics Letters* 100, no. 26, p. 261913.

Wang, T., and R.E. Napolita. 2012. "A Phase-Field Model for Phase Transformations in Glass-Forming Alloys." *Metallurgical and Materials Transactions A* 43, no. 8, pp. 2662–8.

Wang, W., S. Luo, and M. Zhu. 2016a. "Numerical Simulation of Three-Dimensional Dendritic Growth of Alloy: Part I-Model Development and Test." *Metallurgical and Materials Transactions A* 47, no. 3, pp. 1339–54.

Wang, W., S. Luo, and M. Zhu. 2016b. "Numerical Simulation of Three-Dimensional Dendritic Growth of Alloy: Part II-Model Application to Fe-0.82WtPctC Alloy." *Metallurgical and Materials Transactions A* 47, no. 3, pp. 1355–66.

Wang, W.H. 2012. "Metallic Glasses: Family Traits." *Nature Materials* 11, no. 4, pp. 275–6.

Wang, W.H., C. Dong, and C.H. Shek. 2004. "Bulk Metallic Glasses." *Materials Science and Engineering: R: Reports* 44, no. 2–3, pp. 45–89.

Wang, X.D., S. Aryal, C. Zhong, W.Y. Ching, H.W. Sheng, H. Zhang, D.H. Zhang, Q.P. Cao, and J.Z. Jiang. 2015. "Atomic Picture of Elastic Deformation in a Metallic Glass." *Scientific Reports* 5, p. 9184.

Wang, X., P. Lu, N. Dai, Y. Li, C. Liao, and Q. Zheng. 2007. "Noncrystalline Micromachining of Amorphous Alloys using Femtosecond Laser Pulses." *Materials Letters* 61, no. 21, pp. 4290–3.

Wang, X., and X. Xu. 2001. "Molecular Dynamics Simulation of Heat Transfer and Phase Change During Laser Material Interaction." *Journal of Heat Transfer* 124, no. 2, pp. 265–74.

Wang, X., S. Xu, S. Zhou, W. Xu, M. Leary, P. Choong, M. Qian, M. Brandt, and Y.M. Xie. 2016c. "Topological Design and Additive Manufacturing of Porous Metals for Bone Scaffolds and Orthopaedic Implants: A Review." *Biomaterials* 83, pp. 127–41.

Wang, Y., Z. Guo, R. Ma, G. Hao, Y. Zhang, J. Lin, and M. Sui. 2014b. "Investigation of the Microcrack Evolution in a Ti-Based Bulk Metallic Glass Matrix Composite." *Progress in Natural Science: Materials International* 24, no. 2, pp. 121–7.

Wang, Y., W. Jiang, W. Bao, and D. Srolovitz. 2015. "Sharp Interface Model for Solid-State Dewetting Problems with Weakly Anisotropic Surface Energies." *Physical Review* B 91, no. 4, p. 045303.

Wang, Y.-C., C.-Y. Wu, J.P. Chu, and P.K. Liaw. 2010c. "Indentation Behavior of Zr-Based Metallic-Glass Films via Molecular-Dynamics Simulations." *Metallurgical and Materials Transactions A* 41, no. 11, pp. 3010–7.

Wang, Y.S., G.J. Hao, Y. Zhang, J.P. Lin, L. Song, and J.W. Qiao. 2014c. "The Role of the Interface in a Ti-Based Metallic Glass Matrix Composite with in situ Dendrite Reinforcement." *Surface and Interface Analysis* 46, no. 5, pp. 293–6.

Wang, Z., K. Georgarakis, K.S. Nakayama, Y. Li, A.A. Tsarkov, D. Dunida, D.V. Louzguine – Luzgin, and A.R. Yavari. 2016d. "Microstructure and Mechanical Behavior of Metallic Glass Fiber-reinforced Al Alloy Matrix Composites." *Scientific Reports* 6, p. 24384.

Wang, Z.-J., S. Lao, H.-W. Song, W.D. Deng, and W.Y. Li. 2014d. "Simulation of Microstructure during Laser Rapid Forming Solidification Based on Cellular Automaton." *Mathematical Problems in Engineering* 2014, p. 9.

Warnke, P.H., T. Douglas, P. Wollny, E. Sherry, M. Steiner, S. Galonska, S.T., I.N. Springer, J. Wiltfang, and S. Sivananthan. 2008. "Rapid Prototyping: Porous Titanium Alloy Scaffolds Produced by Selective Laser Melting for Bone Tissue Engineering." *Tissue Engineering Part C: Methods* 15, no. 2, pp. 115–24.

Wei, L., X. Lin, M. Wang, W. Huang. n.d. "Low Artificial Anisotropy Cellular Automaton Model and its Applications to the Cell-to-dendrite Transition in Directional Solidification." *Materials Discovery* 3, pp. 17–28.

Wei, S., F. Yang, J. Bednarcik, I. Kaban, O. Shuleshova, A. Meyer, and R. Busch. 2013. "Liquid-Liquid Transition in a Strong Bulk Metallic Glass-Forming Liquid." *Nature Communications* 4, p. 2083.

Weinberg, M.C., D.R. Uhlmann, and E.D. Zanotto. 1989. "'Nose Method' of Calculating Critical Cooling Rates for Glass Formation." *Journal of the American Ceramic Society* 72, no. 11, pp. 2054–8.

Weinberg, M.C., B.J. Zellinski, D.R. Uhlmann, and E.D. Zanotto. 1990. "Critical Cooling Rate Calculations for Glass Formation." *Journal of Non-Crystalline Solids* 123, no. 1, pp. 90–6.

Welk, B.A., H.L. Frazer, V. Dixit, T.M.A. Williams, and M.A. Gibson. 2014. "Phase Selection in a Laser Surface Melted Zr-Cu-Ni-Al-Nb Alloy." *Metallurgical and Materials Transactions B* 45, no. 2, pp. 547–54.

Welk, B.A., M.A. Gibson, and H.L. Fraser. 2016. "A Combinatorial Approach to the Investigation of Metal Systems that Form Both Bulk Metallic Glasses and High Entropy Alloys." *JOM* 68, no. 3, pp. 1021–6.

Wen, P., Z.F. Zhao, M.X. Pan, and W.H. Wang. 2010. "Mechanical Relaxation in Supercooled Liquids of Bulk Metallic Glasses." *Physica Status Solidi (a)* 207, no. 12, pp. 2693–703.

Westhoff, D., J.J. van Franeker, T. Brereton, D.P. Kroese, R.A.J. Janssen, and V. Schmidt, 2015. "Stochastic Modeling and Predictive Simulations for the Microstructure of Organic Semiconductor Films Processed with Different Spin Coating Velocities." *Modelling and Simulation in Materials Science and Engineering* 23, no. 4, p. 045003.

Wong, K.V., and A. Hernandez. 2012. "A Review of Additive Manufacturing." *ISRN Mechanical Engineering* 2012, p. 10.

Wood, W.W., and J.D. Jacobson. 1957. "Preliminary Results from a Recalculation of the Monte Carlo Equation of State of Hard Spheres." *The Journal of Chemical Physics* 27, no. 5, pp. 1207–8.

Wu, C.-Y., Y.-C. Wang, and C. Chen. 2015. "Indentation Properties of Cu-Zr-Al Metallic-Glass Thin Films at Elevated Temperatures via Molecular Dynamics Simulation." *Computational Materials Science* 102, pp. 234–42.

Wu, D., K. Song, C. Cao, R. Li, G. Wang, Y. Wu, F. Wan, et al. 2015a. "Deformation-Induced Martensitic Transformation in Cu-Zr-Zn Bulk Metallic Glass Composites." *Metals* 5, no. 4, p. 2134.

Wu, D.Y., K.K. Song, P. Gargarella, C.D. Cao, R. Li, I. Kaban, and J. Eckert. 2016. "Glass-forming Ability, Thermal Stability of B2 CuZr Phase, and Crystallization Kinetics for Rapidly Solidified Cu-Zr-Zn Alloys." *Journal of Alloys and Compounds* 664, pp. 99–108.

Wu, F.F., Z.F. Zhang, X.S. Mao, A. Peker, and J. Eckert. 2007. "Effect of Annealing on the Mechanical Properties and Fracture Mechanisms of a $Zr_{56.2}Ti_{13.8}Nb_{5.0}Cu_{6.9}Ni_{5.6}Be_{12.5}$ Bulk-Metallic-Glass Composite." *Physical Review* B 75, no. 13, p. 134201.

Wu, F.F., Z.F. Zhang, A. Peker, S.X. Mao, J. Das, and J. Eckert. 2006. "Strength Asymmetry of Ductile Dendrites Reinforced Zr- and Ti-Based Composites." *Journal of Materials Research* 21, no. 09, pp. 2331–6.

Wu, J., Y. Pan, X. Li, and X. Wang. 2014a. "New Insight on Glass-Forming Ability and Designing Cu-Based Bulk Metallic Glasses: The Solidification Range Perspective." *Materials & Design* 61, pp. 199–202.

Wu, W., L. Jiang, H. Jiang, X. Pan, Z. Cao, D. Deng, T. Wang, and T. Li. 2015b. "Phase Evolution and Properties of Al2CrFeNiMo x High-Entropy Alloys Coatings by Laser Cladding." *Journal of Thermal Spray Technology* 24, no. 7, pp. 1333–40.

Wu, X., and Y. Hong. 2001. "Fe-Based Thick Amorphous-Alloy Coating by Laser Cladding." *Surface and Coatings Technology* 141, no. 2–3, pp. 141–4.

Wu, X., B. Xu, and Y. Hong. 2002. "Synthesis of Thick Ni66Cr5Mo4Zr6P15B4 Amorphous Alloy Coating and Large Glass-Forming Ability by Laser Cladding." *Materials Letters* 56, no. 5, pp. 838–41.

Wu, Y., H. Wang, X.J. Liu, X.H. Chen, X.D. Hui, Y. Zhang, and Z.P. Lu. 2014b. "Designing Bulk Metallic Glass Composites with Enhanced Formability and Plasticity." *Journal of Materials Science & Technology* 30, no. 6, pp. 566–75.

Wu, Y., Y. Xiao, G. Chen, C.T. Liu, and Z. Lu. 2010. "Bulk Metallic Glass Composites with Transformation-Mediated Work-Hardening and Ductility." *Advanced Materials* 22, no. 25, pp. 2770–3.

Xi, X.K., D.Q. Zhao, M.X. Pan, W.H. Wang, Y. Wu, and J.J. Lewandowski. 2005. "Fracture of Brittle Metallic Glasses: Brittleness or Plasticity." *Physical Review Letters* 94, no. 12, p. 125510.

Xing, L.Q., J. Eckert, W. Loser, and L. Schultz. 1998. "Effect of Cooling Rate on the Precipitation of Quasicrystals from the Zr-Cu-Al-Ni-Ti Amorphous Alloy." *Applied Physics Letters* 73, no. 15, pp. 2110–2.

Xing, L.Q., J. Eckert, W. Loser, and L. Schultz. 1999. "High-Strength Materials Produced by Precipitation of Icosahedral Quasicrystals in Bulk Zr-Ti-Cu-Ni-Al Amorphous Alloys." *Applied Physics Letters* 74, no. 5, pp. 664–6.

Xu, D., G. Duan, and W.L. Johnson. 2004. "Unusual Glass-Forming Ability of Bulk Amorphous Alloys Based on Ordinary Metal Copper." *Physical Review Letters* 92, no. 24, p. 245504.

Xu, J., U. Ramamurty, and E. Ma. 2010. "The Fracture Toughness of Bulk Metallic Glasses." *JOM* 62, no. 4, pp. 10–18.

Xue, Y.F., H.N. Cai, L. Wang, F.C. Wang, and H.F. Zhang. 2007. "Strength-Improved Zr-Based Metallic Glass/Porous Tungsten Phase Composite by Hydrostatic Extrusion." *Applied Physics Letters* 90, no. 8, p. 081901.

Xue, Y.F., H.N. Cai, L. Wang, F.C. Wang, H.F. Zhang, and Z.Q. Hu. 2008. "Deformation and Failure Behavior of a Hydrostatically Extruded Zr38Ti-17Cu10.5Co12Be22.5 Bulk Metallic Glass/Porous Tungsten Phase Composite Under Dynamic Compression." *Composites Science and Technology* 68, no. 15–16, pp. 3396–400.

Yan, M., S. Kohara, J.Q. Wang, K. Nogita, G.B. Schaffer, and M. Qian. 2011. "The Influence of Topological Structure on Bulk Glass Formation in Al-Based Metallic Glasses." *Scripta Materialia* 65, no. 9, pp. 755–8.

Yang, B.J., J.H. Yao, Y.S. Chao, J.Q. Wang, and E. Ma. 2010. "Developing Aluminum-Based Bulk Metallic Glasses." *Philosophical Magazine* 90, no. 23, pp. 3215–31.

Yang, B.J., J.H. Yao, J. Zhang, H.W. Yang, J.Q. Wang, and E. Ma. 2009. "Al-Rich Bulk Metallic Glasses with Plasticity and Ultrahigh Specific Strength." *Scripta Materialia* 61, no. 4, pp. 423–6.

Yang, G., X. Lin, F. Liu, Q. Hu, L. Ma, J. Li, and W. Huang. 2012. "Laser Solid Forming Zr-Based Bulk Metallic Glass." *Intermetallics* 22, pp. 110–5.

Yang, H., K.Y. Lim, and Y. Li. 2010. "Multiple Maxima in Glass-Forming Ability in Al-Zr-Ni System." *Journal of Alloys and Compounds* 489, no. 1, pp. 183–7.

Yang, M.H., J.H. Li, and B.X. Liu. 2016. "Proposed Correlation of Structure Network Inherited from Producing Techniques and Deformation Behavior for Ni-Ti-Mo Metallic Glasses via Atomistic Simulations." *Scientific Reports* 6, p. 29722.

Yang, Y., and C.T. Liu. 2012. "Size Effect on Stability of Shear-Band Propagation in Bulk Metallic Glasses: An Overview." *Journal of Materials Science* 47, no. 1, pp. 55–67.

Yao, K.F., F. Ruan, Y.Q. Yang, and N. Chen. 2006. "Superductile Bulk Metallic Glass." *Applied Physics Letters* 88, no. 12, p. 122106.

Yap, C.Y., C.K. Chua, Z.L. Dong, Z.H. Liu, D.Q. Zhang, L.E. Loh, and S.L. Sing. 2015. "Review of Selective Laser Melting: Materials and Applications." *Applied Physics Reviews* 2, no. 4, p. 041101.

Yi, J., W. Xu, X. Xiong, L. Kong, M. Ferry, and J. 2016. "Glass-Forming Ability and Crystallization Behavior of Al86Ni9La5 Metallic Glass with Si Addition." *Advanced Engineering Materials* 18, no. 6, pp. 972–7.

Yoshioka, H., K. Asami, A. Kawashima, and K. Hashimoto. 1987. "Laser-Processed Corrosion-Resistant Amorphous Ni-Cr-P-B Surface Alloys on a Mild Steel." *Corrosion Science* 27, no. 9, pp. 981–95.

Yue, T.M. and Y.P. Su. 2008. "Laser Cladding of SiC Reinforced Zr65Al7.5Ni10Cu17.5 Amorphous Coating on Magnesium Substrate." *Applied Surface Science* 255, no. 5, Part 1, pp. 1692–8.

Yue, T.M., Y.P. Su, and H.O. Yang. 2007. "Laser Cladding of Zr65Al7.5Ni10Cu17.5 Amorphous Alloy on Magnesium." *Materials Letters* 61, no. 1, pp. 209–12.

Yves-Christian, H., W. Jan, M. Wilhelm, W. Konrad, and P. Reinhart. 2010. "Net Shaped High Performance Oxide Ceramic Parts by Selective Laser Melting." *Physics Procedia* 5, pp. 587–94.

Zaeem, M.A., H. Yin, and S.D. Felicelli. 2012. "Comparison of Cellular Automaton and Phase Field Models to Simulate Dendrite Growth in Hexagonal Crystals." *Journal of Materials Science & Technology* 28, no. 2, pp. 137–46.

Zeeb, C.N. 2002. *Performance and Accuracy Enhancements of Radiative Heat Transfer Modeling via Monte Carlo* [PhD Thesis]. Fort Collins, CO: Colorado State University.

Zekovic, S., R. Dwivedi, and R. Kovacevic. 2005. Thermo-structural finite element analysis of direct laser metal deposited thin-walled structures. In *Proceedings of SFF Symposium*, Austin, TX.

Zhai, H., H. Wang, and F. Liu. 2016. "A Strategy for Designing Bulk Metallic Glass Composites with Excellent Work-Hardening and Large Tensile Ductility." *Journal of Alloys and Compounds* 685, pp. 322–30.

Zhang, F., M. Ji, X.W. Fang, Y. Sun, C.Z. Wang, M.I. Mendelev, M.J., R.E. Napolitano, and K.M. Ho. 2014a. "Composition-Dependent Stability of the Medium-Range Order Responsible for Metallic Glass Formation." *Acta Materialia* 81, pp. 337–44.

Zhang, H., Y. Pan, and Y. He. 2011a. "Effects of Annealing on the Microstructure and Properties of 6FeNiCoCrAlTiSi High-Entropy Alloy Coating Prepared by Laser Cladding." *Journal of Thermal Spray Technology* 20, no. 5, pp. 1049–55.

Zhang, H., Y. Pan, and Y.-Z. He. 2011b. "Synthesis and Characterization of FeCoNiCrCu High-Entropy Alloy Coating by Laser Cladding." *Materials and Design* 32, no. 4, pp. 1910–5.

Zhang, H., Y. Pan, Y. He, and H. Jiao. 2011c. "Microstructure and Properties of 6FeNiCoSiCrAlTi High-Entropy Alloy Coating Prepared by Laser Cladding." *Applied Surface Science* 257, no. 6, pp. 2259–63.

Zhang, H.F., H. Li, A.M. Wang, H.M. Fu, B.Z. Ding, and Z.Q. Hu. 2009. "Synthesis and Characteristics of 80 vol.% Tungsten (W) Fibre/Zr based Metallic Glass Composite." *Intermetallics* 17, no. 12, pp. 1070–7.

Zhang, J., F. Liou, W. Seufzer, J. Newkirk, Z. Fan, H. Liu, and T.E. Sparks. 2013a. Probabilistic simulation of solidification microstructure evolution during laser-based metal deposition. In *Proceedings of 2013 Annual International Solid Freeform Fabrication Symposium-An Additive Manufacturing Conference,* University of Texas at Austin, TX.

Zhang, K., M. Wang, S. Papanikolaou, Y. Liu, J. Schroers, M.D. Shattuck, and C. O'Heru. 2013b. "Computational Studies of the Glass-Forming Ability of Model Bulk Metallic Glasses." *The Journal of Chemical Physics* 139, no. 12, p. 124503.

Zhang, L.C., D. Klemm, J. Eckert, Y.L. Hao, and T.B. Sercombe. 2011d. "Manufacture by Selective Laser Melting and Mechanical Behavior of a Biomedical Ti-24Nb-4Zr-8Sn Alloy." *Scripta Materialia* 65, no. 1, pp. 21–24.

Zhang, M.X., and P.M. Kelly. 2005a. "Edge-to-Edge Matching and Its Applications: Part I. Application to the Simple HCP/BCC System." *Acta Materialia* 53, no. 4, pp. 1073–84.

Zhang, M.X., and P.M. Kelly. 2005b. "Edge-to-Edge Matching Model for Predicting Orientation Relationships and Habit Planes—The Improvements." *Scripta Materialia* 52, no. 10, pp. 963–8.

Zhang, P., H. Yan, C. Yao, Z. Li, Z. Yu, and P. Xu. 2011e. "Synthesis of Fe-Ni-B-Si-Nb Amorphous and Crystalline Composite Coatings by Laser Cladding and Remelting." *Surface and Coatings Technology* 206, no. 6, pp. 1229–36.

Zhang, Q., Haifeng Zhang, Zhengwang Zhu, and Zhuangqi Hu. 2005. "Formation of High Strength In-Situ Bulk Metallic Glass Composite with Enhanced Plasticity in Cu50Zr47:5Ti2:5 Alloy." *Materials Transactions* 46, no. 3, pp. 730–3.

Zhang, R., T. Jing, W. Jie, and B. Liu. 2006. "Phase-Field Simulation of Solidification in Multicomponent Alloys Coupled With Thermodynamic and Diffusion Mobility Databases." *Acta Materialia* 54, no. 8, pp. 2235–9.

Zhang, T., H.Y. Ye, J.Y. Shi, H.J. Yang, and J.W. Qiao. 2014b. "Dendrite Size Dependence of Tensile Plasticity of in situ Ti-Based Metallic Glass Matrix Composites." *Journal of Alloys and Compounds* 583, pp. 593–7.

Zhang, X., J. Zhao, H. Jiang, and M. Zhu. 2012. "A Three-Dimensional Cellular Automaton Model for Dendritic Growth in Multi-component Alloys." *Acta Materialia* 60, no. 5, pp. 2249–57.

Zhang, Y., and A.L. Greer. 2007. "Correlations for Predicting Plasticity or Brittleness of Metallic Glasses." *Journal of Alloys and Compounds* 434–435, pp. 2–5.

Zhang, Y., X. Lin, L. Wang, L. Wei, F. Liu, and W. Huang. 2015. "Microstructural Analysis of Zr55Cu30Al10Ni5 Bulk Metallic Glasses by Laser Surface Remelting and Laser Solid Forming." *Intermetallics* 66, pp. 22–30.

Zheng, B., Y. Zhou, J.E. Smugeresky, and E.J. Lavernia. 2009. "Processing and Behavior of Fe-Based Metallic Glass Components via Laser-Engineered Net Shaping." *Metallurgical and Materials Transactions A* 40, no. 5, pp. 1235–45.

Zheng, B., Y. Zhou, J.E. Smugeresky, J.M. Schoenung, and E.J. Lavernia. 2008. "Thermal Behavior and Microstructural Evolution during Laser Deposition with Laser-Engineered Net Shaping: Part I. Numerical Calculations." *Metallurgical and Materials Transactions* A 39, no. 9, pp. 2228–36.

Zheng, G.-P. 2012. "A Density Functional Theory Study on the Deformation Behaviors of Fe-Si-B Metallic Glasses." *International Journal of Molecular Sciences* 13, no. 8, pp. 10401–9.

Zhnag, Z.X., C.L. Dai, and J. Xu. 2009. "Complete Composition Tunability of Cu (Ni)–Ti–Zr Alloys for Bulk Metallic Glass Formation." *Journal of Materials Science and Technology* 25, no. 1, pp. 39–47.

Zhou, M., A.J. Rosakis, and G. Ravichandran. 1998. "On the Growth of Shear Bands and Failure-mode Transition in Prenotched Plates: A Comparison of Singly and Doubly Notched Specimens." *International Journal of Plasticity* 14, no. 4, pp. 435–51.

Zhou, X., H. Zhang, G. Wang, X. Bai, X. Fu, and J. Zhao. 2016. "Simulation of Microstructure Evolution during Hybrid Deposition and Micro-rolling Process." *Journal of Materials Science* 51, no. 14, pp. 6735–49.

Zhu, D., W. Zhou, C.S. Ray, and D.E. Day. 2007a. "Method for Estimating the Critical Cooling Rate for Glass Formation from Isothermal TTT Data." *Key Engineering Materials* 336–338, pp. 1874–7

Zhu, M.F., S.Y. Lee, and C.P. Hong. 2004. "Modified Cellular Automaton Model for the Prediction of Dendritic Growth with Melt Convection." *Physical Review* E 69, no. 6, p. 061610.

Zhu, M.F. and D.M. Stefanescu. 2007. "Virtual Front Tracking Model for the Quantitative Modeling of Dendritic Growth in Solidification of Alloys." *Acta Materialia* 55, no. 5, pp. 1741–55.

Zhu, Q., Q. Shiyao, W. Xinhong, and Z. Zengda. 2007b. "Synthesis of Fe-Based Amorphous Composite Coatings with Low Purity Materials by Laser Cladding." *Applied Surface Science* 253, no. 17, pp. 7060–4.

Zhu, Z., C. Yi, T. Shi, Y. Gao, C. Wen, and G. Liao. 2014. "Fabricating Zr-Based Bulk Metallic Glass Microcomponent by Suction Casting Using Silicon Micromold." *Advances in Mechanical Engineering* 6, p. 362484.

Zhu, Z., H. Zhang, Z. Hu, W. Zhang, and A. Inoue. 2010. "Ta-particulate Reinforced Zr-Based Bulk Metallic Glass Matrix Composite with Tensile Plasticity." *Scripta Materialia* 62, no. 5, pp. 278–81.

Zhuang, S., J. Lu, and G. Ravichandran. 2002. "Shock Wave Response of a Zirconium-based Bulk Metallic Glass and Its Composite." *Applied Physics Letters* 80, no. 24, pp. 4522–4.

Zinoviev, A., O. Zinovieva, V. Ploshikhin, V. Romanova, and R. Balokhonov. 2016. "Evolution of Grain Structure During Laser Additive Manufacturing. Simulation by a Cellular Automata Method." *Materials and Design* 106, pp. 321–29.

Zou, J. 1989. *Simulation de la solidification eutectique équiaxe* [PhD thesis No. 774]. Switzerland: Ecole Polytechnique Federale de Lausanne.

Zu, F.-Q. 2015. "Temperature-Induced Liquid-Liquid Transition in Metallic Melts: A Brief Review on the New Physical Phenomenon." *Metals* 5, no. 1, p. 395.

About the Author

Muhammad Musaddique Ali RAFIQUE is a final year PhD student at RMIT University, Melbourne, Australia. He is also a member of the Materials Research Society (MRS), The Minerals, Metals and Materials Society (TMS), Materials Australia (MA), and the American Society for Mechanical Engineers (ASME). Prior to his PhD, he participated in Master in Nanoscience program in Spain and Joint Master Program in Materials Science in Germany and Portugal funded by the European Commission. He also has various practical industry and research experiences in fields such as physical and mechanical metallurgy, metal casting, foundry, welding and joining, sheet metal work, manufacturing processes, materials characterization, modeling and simulation, coding, optimization, scale up, and applications. He has also previously published widely and is lead author of 12 research papers, 3 books, and 2 book chapters. His main area of focus during his doctoral research was bulk metallic glasses and their composites and their additive manufacturing. He has several publications on the topic or related topics and has adequate expertise in the field as manifested by his number of years of experience and citations of the work.

INDEX

OTHER TITLE IN OUR EMERGING MATERIALS COLLECTION

N.M. Ravindra, New Jersey Institute of Technology, *Editor*

- *Black Silicon: Processing, Properties, and Applications* by Nuggehalli M. Ravindra

  Momentum Press offers over 30 collections including Aerospace, Biomedical, Civil, Environmental, Nanomaterials, Geotechnical, and many others. We are a leading book publisher in the field of engineering, mathematics, health, and applied sciences.

  Momentum Press is actively seeking collection editors as well as authors. For more information about becoming an MP author or collection editor, please visit http://www.momentumpress.net/contact

Announcing Digital Content Crafted by Librarians

Concise e-books business students need for classroom and research

Momentum Press offers digital content as authoritative treatments of advanced engineering topics by leaders in their field. Hosted on ebrary, MP provides practitioners, researchers, faculty, and students in engineering, science, and industry with innovative electronic content in sensors and controls engineering, advanced energy engineering, manufacturing, and materials science.

Momentum Press offers library-friendly terms:
- *perpetual access for a one-time fee*
- *no subscriptions or access fees required*
- *unlimited concurrent usage permitted*
- *downloadable PDFs provided*
- *free MARC records included*
- *free trials*

The **Momentum Press** digital library is very affordable, with no obligation to buy in future years.

For more information, please visit **www.momentumpress.net/library** or to set up a trial in the US, please contact **mpsales@globalepress.com**.